Ferrite materials for memory applications

by R. Saravanan

In the modern world, the life style of humans is greatly influenced by electronic gadgets. These electronic gadgets need semiconducting and magnetic materials. In particular, the magnetic materials which find applications in almost all such gadgets need to be researched and better understood. Magnetism has diverse applications, from simple "loadstone" to complex DNA sequencing.

The aim of this book is to describe the synthesis and characterization of various nano ferrite materials used for memory applications. It is now well established that materials synthesized in nanometer scale have novel properties compared to their bulk counterparts. The distinct feature of the book is the construction of charge density diagrams of ferrites by using the maximum entropy method (MEM). It is analyzed how the charge density distribution in the ferrite unit cell affects charge related properties.

Ferrite materials for memory applications

by

Dr. R. Saravanan, M.Sc., M.Phil., Ph.D.
Associate Professor & Head
Research Centre and PG Department of Physics
The Madura College (Autonomous)
Madurai - 625 011
India

Published by **Materials Research Forum LLC**
Millersville, PA 17551, USA

Published as part of the book series
Materials Research Foundations
Volume 18 (2017)
ISSN 2471-8890 (Print)
ISSN 2471-8904 (Online)

Print ISBN 978-1-945291-38-8
ePDF ISBN 978-1-945291-39-5

Distributed worldwide by

Materials Research Forum LLC
105 Springdale Lane
Millersville, PA 17551
USA
http://www.mrforum.com/

Manufactured in the United States of America
10 9 8 7 6 5 4 3 2 1

Table of Contents

Preface

Chapter 1   **Introduction** 1

1.   Introduction 3

1.1   Objectives of the present work 4

1.2   Methodologies and software programs employed in this research work 5

1.3   Magnetism and classification of magnetic materials 5

1.4   Spinel ferrites 7

1.5   Types of spinel ferrites 10

1.6   Magnetic interactions in ferrites 10

1.7   Synthesis of ferrite samples 12

1.7.1   Mechanical alloying method 12

1.7.2   Sintering 12

1.7.3   Pre-sintering 12

1.7.4   Final sintering 12

1.7.5   Co-precipitation method 13

1.7.6   Molten – salt low temperature method 13

1.7.7   Synthesis of $Ni_{0.5}Zn_{0.5}Fe_2O_4$ nano ferrite particles by mechanical alloying 14

1.7.8   Synthesis of $(XxYyZz)\ Fe_2O_4$ ferrites samples by co-precipitation method 14

1.7.9   Synthesis of $NiFe_2O_4$ nano ferrite particles 15

1.8   Characterizations techniques 16

1.8.1   Structural studies – X-Ray diffraction 16

1.8.2   Bragg's law 16

1.8.3   Powder diffraction 17

1.8.4   Thermal analysis 18

1.8.5	Morphology studies - Scanning electron microscope (SEM)	18
1.8.6	Elemental analysis and compositional studies – EDX and XRF	18
1.8.6.1	Energy dispersive X-Ray analysis	18
1.8.6.2	X-Ray Fluorescence (XRF)	18
1.8.7	Magnetic studies – Vibrating sample magnetometer	19
1.8.8	Hysteresis loop	19
1.8.9	Optical studies - UV-Visible spectroscopy	20
1.8.10	Dielectric studies – Broad band dielectric spectrometer	21
1.8.11	Polarization	21
1.8.12	Effects of frequency on polarization	22
1.8.13	Dielectric constant	22
1.8.14	Frequency dependence of dielectric constant	23
1.8.15	Dielectric strength	23
1.9	Characterizations of ferrites	23
1.10	Methodologies employed in this research work	24
1.10.1	Rietveld method	25
1.10.2	Theory of Rietveld method	25
1.10.3	Rietveld Refinement	26
1.10.4	Maximum Entropy Method (MEM)	26
1.10.5	Implication of MEM	27
1.10.6	Principle of maximum entropy method (MEM)	27
1.10.7	Theory of MEM	27
1.11	Theories involved in the present research work	29
1.11.1	Estimation of cation distribution	29
1.11.2	Estimation of magnetic properties of ferrites	32
1.11.3	Estimation of optical band gap energy	33
1.11.3.1	Theory	33
1.11.4	Estimation of dielectric properties	35
1.12	Photographs of samples synthesized in this research study	35

References 36

Chapter 2 .... **42**

2. Introduction 43

2.1 Structural characterization of the synthesized samples 43

2.2 Results – Raw powder XRD data 44

2.3 Results – Cation distribution... 46

2.4 Results – Refined powder XRD of synthesized samples.................... 48

2.5 TG/DTA analysis of $Ni_{0.5}Zn_{0.5}Fe_2O_4$ (700 - 1000°C)
and $Ni_{0.8}Zn_{0.2}Fe_2O_4$ samples... 64

2.6 Morphological studies of synthesized samples................................... 67

2.6.1 SEM results 67

2.6.2 EDAX and XRF characterization results 75

2.7 Results – MEM analysis 81

2.8 Results - Magnetic properties 105

2.9 Results - Optical properties.. 111

2.10 Results - Dielectric properties 115

References 119

Chapter 3 **Discussion.. 120**

3.1 Discussion - Structural parameters of synthesized samples 122

3.1.1 Phase identification in the synthesized samples from
raw powder XRD 122

3.1.2 Cation distribution in the unit cell of synthesized samples............... 123

3.1.3 Discussion on refinement of raw XRD 123

3.1.3.1 Experimental lattice parameter of synthesized samples 124

3.1.3.2 Oxygen positional parameter of synthesized samples 125

3.1.3.3 Other structural parameters of synthesized samples......................... 126

3.1.3.4 Theoretical lattice parameter of synthesized samples....................... 128

3.1.3.5 Variation between theoretical and experimental lattice parameter in
$Ni_{0.5}Zn_{0.5}Fe_2O_4$ samples 128

3.2   Discussion on TG/DTA analysis of $Ni_{0.5}Zn_{0.5}Fe_2O_4$ (700 to 1000°C)
            and $Ni_{0.8}Zn_{0.2}Fe_2O_4$ sample 128

3.3   Discussion on morphological studies of synthesized samples........... 129

3.3.1   Discussion on SEM images.. 129

3.3.2   Discussion on EDAX results... 130

3.3.3   Discussion on XRF results.. 131

3.4   Discussion on MEM results 131

3.4.1   Theories regarding various sites interactions in ferrite unit cell........ 131

3.4.2   Results based on theory regarding various site interactions in ferrite
            unit cell 132

3.4.3   MEM results on various sites interactions in ferrite unit cell 133

3.4.3.1 $Ni_{0.5}Zn_{0.5}Fe_2O_4$ samples 134

3.4.3.2 $Ni_{0.8}Zn_{0.2}Fe_2O_4$ sample 135

3.4.3.3 $Mn_{0.4}Zn_{0.6}Fe_2O_4$ and $Co_{0.4}Zn_{0.6}Fe_2O_4$ samples.................................. 135

3.4.3.4 $Ni_{0.53}Cu_{0.12}Zn_{0.35}Fe_2O_4$ and $Mg_{0.2}Cu_{0.3}Zn_{0.5}Fe_2O_4$ samples................ 136

3.4.3.5 $NiFe_2O_4$ sample 136

3.5   Discussion on magnetic properties 136

3.5.1   Saturation magnetization of the synthesized samples from
            the hysteresis curve 137

3.5.2   Coercivity of the synthesized samples from the hysteresis curve...... 138

3.5.3   Evaluation of magnetic moment from saturation magnetization
            and cation distribution data 140

3.6   Discussion on optical properties 142

3.7   Discussion on dielectric properties 143

3.7.1   Variation of dielectric constant with frequency of
            $Ni_{0.5}Zn_{0.5}Fe_2O_4$ samples 143

3.7.2   Variation of tanδ with frequency of $Ni_{0.5}Zn_{0.5}Fe_2O_4$ samples............ 144

3.7.3   Variation of AC conductivity with frequency of
            $Ni_{0.5}Zn_{0.5}Fe_2O_4$ samples 145

3.7.4   Variation of dielectric constant with frequency of $NiFe_2O_4$ samples 145

References   147

Chapter 4   **Conclusions** **153**

4   Summary 153

4.1 $Ni_{0.5}Zn_{0.5}Fe_2O_4$ samples sintered from 700-1400°C 153

4.2 $Ni_{0.8}Zn_{0.2}Fe_2O_4$ sample 155

4.3 $Mn_{0.4}Zn_{0.6}Fe_2O_4$ and $Co_{0.4}Zn_{0.6}Fe_2O_4$ samples................................... 155

4.4 $Ni_{0.53}Cu_{0.12}Zn_{0.35}Fe_2O_4$ and $Mg_{0.2}Cu_{0.3}Zn_{0.5}Fe_2O_4$ samples................ 155

4.5 $NiFe_2O_4$ sample 156

Keywords   157

About the author 158

Preface

In the modern world, the life style of humans is greatly influenced by electronic gadgets. These electronic gadgets possess semiconducting and magnetic materials. In particular, the magnetic materials which find applications in almost all such gadgets need to be further researched. Magnetism has diverse applications, from simple "loadstone" to a complex DNA sequencing. The subject magnetism is an inter-disciplinary field in which not only physicists but also materials scientists, chemists, etc do their research.

The book discusses the synthesis and characterization of various ferrite materials used for memory applications. It is now well established that materials synthesized in nanometer scale have novel properties compared to their bulk counterpart. The distinct feature of the book is the construction of charge density of ferrites by deploying maximum entropy method (MEM). This charge density gives the distribution of charges in the ferrite unit cell, which is analyzed for charge related properties.

Chapter 1 depicts the objectives of the present research work. A brief introduction to magnetism and spinel ferrites is given. It also highlights the various synthesis method employed in the present research work for preparing nano ferrites. The principle behind the various characterization techniques such as, powder X-ray diffraction, scanning electron microscopy, vibrating sample magnetometry, UV Visible spectrometry and broad band dielectric spectrometry which have been used for analyzing the structural properties of the crystal systems, morphological properties, magnetic properties, optical properties and dielectric properties of the materials respectively are discussed. This chapter presents the basics of PXRD profile fitting technique and the methodology adopted for calculating the experimental charge densities. It also gives an account on the theoretical aspects involved in estimating various properties of ferrites discussed.

Chapter 2 displays the results of the powder X-ray diffraction, scanning electron microscope, energy dispersive X-Ray analysis, X-Ray fluorescence, magnetic, optical and dielectric characterizations carried out on the $NiFe_2O_4$, $Ni_{0.5}Zn_{0.5}Fe_2O_4$, $Ni_{0.8}Zn_{0.2}Fe_2O_4$, $Mn_{0.4}Zn_{0.6}Fe_2O_4$, $Co_{0.4}Zn_{0.6}Fe_2O_4$, $Ni_{0.53}Cu_{0.12}Zn_{0.35}Fe_2O_4$ and $Mg_{0.2}Cu_{0.3}Zn_{0.5}Fe_2O_4$ ferrite samples synthesized in this research work. The results of fitted PXRD profiles for all the prepared nano ferrite materials are reported. Particle size evaluation from SEM, crystallite sizes from powder X-ray profile, elemental compositional analyses of the synthesized nano ferrite materials are also presented. Electron density distribution studies are carried out using maximum entropy method. The results of electron density distribution studies are presented in the form of three, two and one dimensional electron density maps.

Chapter 3 brings out a detailed discussion on the experimental lattice parameters, theoretical lattice parameters and oxygen positional parameters, radii of tetrahedral and octahedral sites evaluated from the results of powder XRD profile fitting and from the knowledge of cation distribution in the ferrite unit cell. Discussion on the mid bond electron density between tetrahedral and octahedral sites evaluated from maximum entropy method is presented and these results are compared with hitherto existing theoretical results. Magnetic properties, optical properties and dielectric properties of the synthesized ferrite samples are also discussed in this chapter.

Chapter 4 reveals the conclusion of the findings of the reported work.

This book includes research findings previously published by the author in the following research publications.

1. Synthesis and characterization of some ferrite nanoparticles prepared by co-precipitation method- Y. B. Kannan, R. Saravanan, N. Srinivasan, K. Praveena, K. Sadhana., *Journal of Materials Science: Materials in Electronics*, Springer Publication, 27(11), 12000 – 12008, 2016.
2. Study of various sites interactions using maximum entropy method on mechanically alloyed $Ni_{0.5}Zn_{0.5}Fe_2O_4$ nano ferrite particles sintered from 1100°C to 1400°C- R. Saravanan, Y.B. Kannan, N. Srinivasan, I. Ismail. *Journal of Superconductivity and Novel Magnetism*, Springer Publication, 30(2), 407-417, 2017.
3. Sintering effect on structural, magnetic and optical properties of $Ni_{0.5}Zn_{0.5}Fe_2O_4$ ferrite nano particles-Y.B. Kannan, R. Saravanan, N. Srinivasan, I. Ismail. *Journal of Magnetism and Magnetic Materials,* Elsevier, 423, 217-225, 2017.
4. Structural, magnetic and optical characterization of $Ni_{0.8}Zn_{0.2}Fe_2O_4$ nano particles prepared by co-precipitation method- Y.B. Kannan, R. Saravanan, N. Srinivasan, K. Praveena, K. Sadhana. *Physica B*, Elsevier, 502, 181-186, 2016.
5. Structural, Magnetic, Optical, and MEM Studies on Co-precipitated $X_{0.4}Zn_{0.6}Fe_2O_4$ (X = Co, Mn) Nanoferrite Particles-Y.B. Kannan, R. Saravanan, N. Srinivasan, K. Praveena, K. Sadhana. *Journal of Superconductivity and Novel Magnetism*, Springer Publication, DOI: 10.1007/s10948-017-4081-x.
6. Effect of sintering on dieletric and AC conductivity properties of $Ni_{0.5}Zn_{0.5}Fe_2O_4$ nano ferrite particles.- Y.B.Kannan, R.Saravanan, N.Srinivasan, I.Ismail. *Journal of the Australian ceramic society*, Springer Publication, DOI: 10.1007/s41779-017-0069-z.

Chapter 1

Introduction

Abstract

This chapter depicts the objectives of the present research work. A brief introduction to magnetism and spinel ferrites is given. It also highlights the various synthesis methods employed for preparing nano ferrites. The principles behind the various characterization techniques such as, powder X-ray diffraction, scanning electron microscopy, vibrating sample magnetometry, UV Visible spectrometry and broad band dielectric spectrometry which have been used for analyzing the structural properties of the crystal systems, morphological properties, magnetic properties, optical properties and dielectric properties of the materials respectively are discussed. This chapter presents the basics of PXRD profile fitting technique and the methodology adopted for calculating the experimental charge densities. It also gives an account on the theoretical aspects involved in estimating various properties of the ferrites.

Keywords

Ferrite materials, Magnetic, $NiFe_2O_4$, $Ni_{0.8}Zn_{0.2}Fe_2O_4$, $Mn_{0.4}Zn_{0.6}Fe_2O_4$, $Co_{0.4}Zn_{0.6}Fe_2O_4$, $Mg_{0.2}Cu_{0.3}Zn_{0.5}Fe_2O_4$, $Ni_{0.53}Cu_{0.12}Zn_{0.35}Fe_2O_4$, Basic Concepts, Characterization Techniques

Contents

1.   Introduction...3

1.1   Objectives of the present work...4

1.2   Methodologies and software programs employed in
      this research work..5

1.3   Magnetism and classification of magnetic materials............5

1.4   Spinel ferrites 7

1.5   Types of spinel ferrites 10

1.6   Magnetic interactions in ferrites 10

1.7   Synthesis of ferrite samples 12

1.7.1 Mechanical alloying method 12

1.7.2 Sintering...12

1.7.3 Pre-sintering 12

1.7.4 Final sintering 12

1.7.5 Co-precipitation method 13

1.7.6 Molten-salt low temperature method 13

1.7.7 Synthesis of $Ni_{0.5}Zn_{0.5}Fe_2O_4$ nano ferrite particles by
        mechanical alloying..13

1.7.8 Synthesis of $(X_xY_yZ_z)$ Fe_2O_4 ferrites samples by
        co-precipitation method 14

1.7.9 Synthesis of $NiFe_2O_4$ nano ferrite particles...................................15

1.8   Characterizations techniques 16

1.8.1 Structural studies – X-Ray diffraction 16

1.8.2 Bragg's law 16

1.8.3 Powder diffraction 17

1.8.4 Thermal analysis...18

1.8.5 Morphology studies - Scanning electron microscope (SEM) 18

1.8.6 Elemental analysis and compositional studies – EDX and XRF........18

1.8.6.1   Energy dispersive X-ray analysis 18

1.8.6.2   X-ray fluorescence (XRF) 18

1.8.7 Magnetic studies – Vibrating sample magnetometer..........................19

1.8.8 Hysteresis loop 19

1.8.9 Optical studies - UV-visible spectroscopy 20

1.8.10 Dielectric studies – Broad band dielectric spectrometer 21

1.8.11 Polarization 21

1.8.12 Effects of frequency on polarization 22

1.8.13 Dielectric constant 22

1.8.14 Frequency dependence of dielectric constant 23

1.8.15 Dielectric strength..23

1.9   Characterizations of ferrites..23

1.10   Methodologies employed in this research work............................24

1.10.1 Rietveld method 25

1.10.2 Theory of Rietveld method 25

1.10.3 Rietveld refinement 26

1.10.4 Maximum entropy method (MEM) 26

1.10.5 Implication of MEM 27

1.10.6 Principle of maximum entropy method (MEM)........................27

1.10.7 Theory of MEM 27

1.11   Theories involved in the present research work............................29

1.11.1 Estimation of cation distribution 29

1.11.2 Estimation of magnetic properties of ferrites 32

1.11.3 Estimation of optical band gap energy 33

1.11.3.1   Theory..33

1.11.4 Estimation of dielectric properties...35

1.12   Photographs of samples synthesized in this research study 35

References 36

1.   Introduction

Ferrites fascinated the scientists and engineers since the beginning of the 20th century around the world as non-metallic ferromagnetic materials, substituting the till metals which have undesirable consequences with increase in frequency. Ferrites continue to be the most hunted materials in the fabrication of electronic devices due to high resistivity, low eddy current and dielectric losses. Spinel ferrites have wide range of applications in the biomedical field, memory devices and electric motors. Magnetic properties of ferrite are interpreted by Neel model (Neel, 1948) with two interpenetrating sub-lattices, with oppositely directed and unequal spin magnetic moments. The properties of ferrites depend on the preparation methods, substitution and composition of magnetic/non-magnetic atoms, microstructures, etc.

With the development in the materials processing technology, production of materials in the nano regime is feasible. The significant impact of particles in nano regime is that the properties are completely different yet useful than that of its bulk counterpart. The change in the properties at nano scale is attributed to the increase in the ratio of surface area to volume of the object, change in the curvature shape on the surface and the size of the particles approaching the characteristics length scales. Research institutes around the world are spending lots of human resources and funds to decipher the conundrum how these novel properties at nanometer regime can be utilized for the betterment of human life.

1.1   Objectives of the present work

One can find applications of ferrite materials from a simple permanent magnet to most sophisticated equipment in medical fields (Nikam et al., 2015). Ferrites have an important advantage over other magnetic materials. The oxygen ions in the ferrites insulate the metal ions in the compounds thereby increase the resistivity which leads to decrease in the eddy current losses. High frequency devices, used in several applications, require materials with low eddy current loss and hence, ferrite materials attracted researchers for the last three decades and still continue to do so. Moreover, the magnetic property of ferrites can be "tailored" by varying the distribution of cations over the available tetrahedral (A) site and octahedral (B) site in the unit cell. The cations, in micrometer regime have their "own" preferred sites (A or B sites) of occupation in the unit cell, whereas in nanometer regime, this occupation constraint is not strictly adhered and hence, cation distribution can be "tailored". Methods of preparation and dopants play a major role in bringing down the particle size to nano scale.

Understanding the bond strengths requires an intense knowledge on charge density distribution. Not much work has been reported so far on bond strengths between atoms at the A-site and B-site of ferrites by deploying maximum entropy method (MEM) techniques (Gull and Daniel, 1978) in the unit cell of ferrites. This motivated to carry out intense research on the bond strengths and other characterization on ferrites. The physical and chemical properties of atoms or molecules mostly depend on the charge density distribution, bond strength, etc. Hence, in this research work, the following samples were synthesized and studied for their bond strength between various sites (A-A, A-B and B-B sites) using MEM (Gull and Daniel, 1978).

1. $Ni_{0.5}Zn_{0.5}Fe_2O_4$.

2. $Ni_{0.8}Zn_{0.2}Fe_2O_4$

3. $Mn_{0.4}Zn_{0.6}Fe_2O_4$

4. $Co_{0.4}Zn_{0.6}Fe_2O_4$

5. $Ni_{0.53}Cu_{0.12}Zn_{0.35}Fe_2O_4$

6. $Mg_{0.2}Cu_{0.3}Zn_{0.5}Fe_2O_4$.

7. $NiFe_2O_4$.

Sample (1) was synthesized by mechanical alloying via ball milling and sintering from 700 to 1400°C. Samples (2-6) were synthesized by co-precipitation method followed by sintering at 900°C. Sample (7) was synthesized by the chemical reaction method and calcined at 700°C.

1.2   Methodologies and software programs employed in this research work

In this research work, the Rietveld refinement method (Rietveld, 1969) was employed to refine the raw X-ray diffraction data using the software program JANA 2006 (Petříček et al., 2006) and MEM (Gull and Daniel, 1978) was used to evaluate the mid bond electron density values between the A-A, A-B and B-B site atoms in the unit cell of ferrite. PRIMA (Practice Iterative MEM Analyses) (Izumi and Dilanian, 2002) software, which is a MEM analysis program, was employed to calculate the mid bond electron densities from the X-ray diffraction data and VESTA (Visualization of Electron/Nuclear densities and Structures) (Momma and Izumi, 2006) software was used to plot the 3D, 2D and 1D plot of electron densities in this research work. Grain size of the samples synthesized in the present research work was determined using the software GRAIN (Saravanan R, GRAIN software) written by Dr. R. Saravanan.

1.3   Magnetism and classification of magnetic materials

Our ancestors used "loadstone" for navigation and with the dawn of science, scientist revealed that loadstone is nothing but naturally occurring iron ore called magnetite (Fe_3O_4) and named it as natural magnet. This loadstone possesses a property called magnetism which means that the material attracts certain substances and repels certain substances. Magnetism of an element is due to the movement of electrons around the nucleus. Electrons are integral part of all elements and hence, all elements in the periodic table should exhibit magnetism. Michael Faraday endorsed this thought by stating that not only iron but all elements are magnetic to a greater or lesser extent. The extent of magnetism of an element is governed by Pauli's exclusion principle. We know that the electron has two kinds of motions around the nucleus namely (i) spin and (ii) orbital motion. These two kinds of motions make the electrons behave like a tiny magnet and the resultant magnetic moment depends upon the direction and magnitude of these two motions. According to Pauli's exclusion principle, the direction of the magnetic moment due to the spin of any two electrons in a given orbit lies exactly in the opposite direction leading to cancellation of net spin magnetic moment and the same applies to the orbital magnetic moments also. The resultant magnetic moment is zero for atom which has paired electrons. If such material is placed in an external magnetic field, the frequency of precession of electron orbit is affected slightly and hence, a magnetic field is induced. This induced magnetic field, as per Lenz's law, opposes the cause of it. These kinds of materials are termed as diamagnetic materials (Cullity, 1959). Magnetic susceptibility value, which is the ratio of magnetization and applied field, is negative for diamagnetic materials.

The magnetic moment is finite for atoms which have unpaired electrons and these atoms are classified into two types namely, paramagnetic and ferromagnetic materials. The net magnetic moment value for paramagnetic material is zero and finite respectively in the absence and presence of external magnetic field (Kakani et al., 2004). The paramagnetic atoms have finite value in the presence of external magnetic field because the external field aligns the randomly oriented magnetic moment to the direction of applied field and hence, a finite value of magnetic moment. Once the external field is withdrawn the alignment break down and the net magnetic moment becomes zero. If the temperature of paramagnetic material increases, then the increase in thermal energy disturbs the alignment between resultant magnetic moment direction and that of the external field. Thus, the paramagnetic effect is temperature dependent. The susceptibility value of paramagnetic materials is positive but very small.

In the case of ferromagnetic material, the net magnetic moment value is finite, even in the absence of an external magnetic field, due to the presence of domains. The coupling interactions, which is attributed to the arrangement of electrons, between magnetic moments of adjacent atoms make the magnetic moment to align in a particular direction in a small region called domain. Ferromagnetic materials consist of many such domains where each domain has magnetic moment aligned in different directions. The presence of an external field to these kinds of materials enhances the magnetism by aligning the magnetic moments of various domains in the field direction which makes the susceptibility value of ferromagnetic material to be positive and large. With the increase in temperature, the alignments get disturbed and above a temperature called Curie temperature the ferromagnetic material becomes paramagnetic materials (Jiles, 1998).

In some materials, the coupling interaction between the adjacent atoms is in such a way that it aligns the magnetic moment in opposite direction resulting in null net magnetic moment and those substances are termed as anti-ferromagnetic materials (Jiles, 1998).

Elements or compounds belonging to different types of magnetic materials are listed in Table 1.1.

Table 1.1 *Example for different types of magnetic materials*

Diamagnetic	Paramagnetic	Ferromagnetic	Anti-Ferromagnetic	Ferrimagnetic
Zinc	Gadolinium	Cobalt	Mangenese oxide	Magnetite
Copper	Magnesium	Iron	Cobalt Oxide Nickel	(Fe_3O_4) and
Silver	Lithium	Nickel	Oxide Copper	Maghemite
Gold	Tantalum		chloride	$(\gamma\text{-}Fe_2O_3)$

Interestingly, another type of magnetism known as ferrimagnetism is found in certain compounds such as oxides, known as ferrites from which ferrimagnetism derives its name. In ferrimagnetism, the adjacent atoms having unequal magnitudes of magnetic moments align themselves in opposite direction and hence, these materials have a resultant magnetism. The susceptibility value of ferro, anti-ferro and ferrimagnetic material is positive and very large (Jiles, 1998).

1.4 Spinel ferrites

The term "spinel" is derived from the naturally occurring mineral $Mg^{2+}Al_2^{3+}O_4$ (Magnetite). If Fe is substituted for Al then that kind of spinel is known as ferrite. The divalent Mg ion can be replaced by any divalent magnetic/non-magnetic ions. Ferrite, in general, can be represented by $Me^{2+}Fe_2^{3+}O_4$ where Me = (Ni, Zn, Fe, Mg, Mn, Cr, Co, Cu etc.). Depending upon the structure, ferrites can be classified as spinel, garnets, magneto-plumbite and pervoskite types of ferrites (Viswanathan et al., 1990). Since the present work is on spinel ferrites, the structure of ferrites is explained here.

The unit cell of ferrite contains eight formula units of $(Me^{2+})[Fe_2^{3+}]O_4$ molecules. The parentheses represent the A-site and the square brackets represent the B-site. Spinel ferrite belongs to the space group $Fd\bar{3}m$ $(F_{1/d}^4\bar{3}_{2/m};O_h^7;$ No. 227) and it has cubic crystal system in which oxygen anions occupy the 32e positions (corners of a fcc lattice) and the divalent (Me^{2+}) and trivalent (Fe_2^{3+}) cations occupy the 8a and 16d positions respectively. Due to the fcc packing of oxygen anions, there will be 64 A-sites and 32 B-sites per unit cell and only one-eight of 64 A-sites and half of the 32 B-sites are occupied in a unit cell of ferrite to maintain charge neutrality. Thus, there will be $8Me^{2+}$ cations, 16 Fe_2^{3+} trivalent cations and 32 oxygen anions in a unit cell (Candeia et al., 2006). The oxygen ion radius, 1.32Å (Nikam et al., 2015) is much larger than that of substituted metal ions (Me^{2+}), whose radii generally lie within the range of 0.6 – 0.8Å and hence, the adjacent oxygen ions in the lattice are brought closer and form a close packed face-centered cubic lattice.

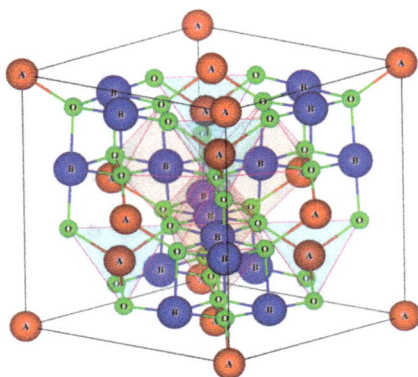

Fig.1.1 *Unit cell of cubic spinel ferrite.*

Fig.1.1 illustrates the unit cell of spinel ferrite, drawn with the VESTA software (Momma and Izumi, 2006), with its A and B-sites. The A-sites are marked as A (red), B-sites are marked as B (Blue) and oxygen ions are represented by O (green). The A-site, B-site and oxygen ions radii are not to scale in Fig.1.1. Four oxygen atoms are surrounding the A-site atom and six oxygen atoms are surrounding the B-site atom. In an ideal closed packed FCC structure of oxygen atoms, ions with radius ≤ 0.3Å occupy A-site and ions with radius ≤ 0.55Å occupy B-site. If Me^{2+} in $(Me^{2+})[Fe_2^{3+}]O_4$ is replaced by any one of the following ions such as Fe, Mg, Mn, Co, Ni, Cu, Zn, Cd, Cr, Sn, etc., having radius greater than 0.6Å then the lattice has to expand to accommodate these ions, leading to deformation in the close packing of oxygen ions (Hemada et al., 2014). The tetrahedral sites are expanded by an equal displacement of the four oxygen ions outwards, along the body diagonal of the cube, still occupying the corners of an expanded regular tetrahedron.

The four oxygen ions of the octahedral sites are shifted in such a way that this oxygen tetrahedron shrinks by the same aount as the first expands. Hence, an oxygen ion moves along the body diagonal from its ideal position and this displacement of an oxygen ion is known as oxygen parameter 'u', as shown in Fig. 1.2. The direction of OA is along the body diagonal direction of a cube and the direction of OB are along the cube edges.

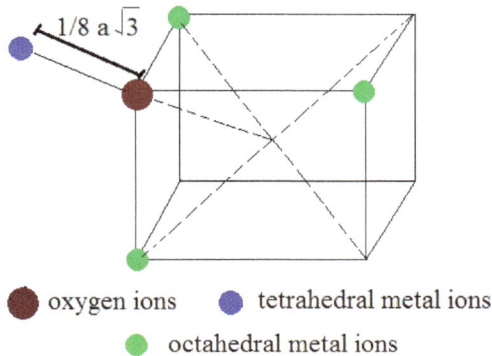

Fig.1.2 *Surroundings of an oxygen ion in spinel ferrite. Each oxygen ion is surrounded by one tetrahedral (A) and three octahedral (B) ions. The direction of OA is the body diagonal direction and that of OB are the cube edges.*

For an ideal FCC structure, the value of 'u' is equal to 0.375Å (Hemada et al., 2014) in which the packing of ions within the unit cell is taken as perfect. But in the actual case, 'u' value is greater than 0.375Å because of expansion in the A-sites, which means the oxygen ions are not located at the exact FCC sub-lattice (Sagar et al., 2014). When the oxygen positional parameter 'u' is equal to 0.250Å, then the origin of the unit cell is considered at B-site (Kurt et al., 1999; Candeia et al., 2006).

The magnetic properties of ferrites (ferrimagnetism) are different from that of ferromagnetic materials like iron etc., in which the magnetic field line up in a single direction with its surroundings. But, in ferrites, the direction of spin magnetic moment of A-site atoms is aligned in one direction and that of B-site atoms is oppositely aligned but with unequal magnitude and as a result, incomplete cancellation of the spin causes a strong net magnetic moment.

At a particular temperature, known as Neel temperature, ferrite materials will become paramagnetic.

1.5   Types of spinel ferrites

Ferrites can be categorized into different types based on composition and on the distribution of cations. Based on composition, ferrites can be classified as (i) simple spinel ferrites, (ii) mixed spinel ferrites and (iii) substitutional spinel ferrites.

If Me^{2+} in $(Me^{2+})[Fe_2^{3+}]O_4$ is substituted by any one of the divalent cation, Me^{2+} = (Ni, Zn, Fe, Mg, Mn, Cr, Co, Cu etc.) then, such ferrites are known as simple spinel ferrites e.g. $NiFe_2O_4$, $ZnFe_2O_4$, $CrFe_2O_4$, $MgFe_2O_4$ etc.

If Me^{2+} in $(Me^{2+})[Fe_2^{3+}]O_4$ is substituted by more than one divalent cations, Me^{2+} = (Ni, Zn, Fe, Mg, Mn, Cr, Co, Cu etc.) with stoichiometric constraint and then such ferrites are known as mixed spinel ferrites e.g. $Ni_{0.5}Zn_{0.5}Fe_2O_4$, $Cr_{0.2}Cu_{0.8}Fe_2O_4$, $Mg_{0.3}Co_{0.7}Fe_2O_4$, $Mg_{0.05}Cu_{0.5}Co_{0.45}Fe_2O_4$ etc.

If Me^{2+} and Fe^{3+} in $(Me^{2+})[Fe_2^{3+}]O_4$ are replaced by other magnetic or nonmagnetic ions (Ni, Zn, Fe, Mg, Mn, Cr, Co, Cu etc.) in the spinel structure, the resulting ferrites are called substitutional ferrites e.g. $(Ni_{0.25}Cu_{0.20}Zn_{0.55})La_xFe_{2-x}O_4$ (x = 0.00, 0.025, 0.050 and 0.075).

Based on the occupation of divalent and trivalent cations over the available tetrahedral and octahedral sites, spinel ferrites can be classified as (i) normal and (ii) inverse spinel ferrites.

In normal spinel ferrite, all the divalent (Me^{2+}) and trivalent (Fe_2^{3+}) cations respectively, occupy the A and B sites in the unit cell of ferrite. A normal spinel is represented by $(Me^{2+})[Fe_2^{3+}]O_4^{2-}$ e.g. $ZnFe_2O_4$.

In inverse spinel ferrite, all the divalent ions (Me^{2+}) occupy the B-sites and trivalent cations (Fe_2^{3+}) distributed equally between the A and B-sites. An inverse spinel is represented by $(Fe^{3+})[Me^{2+}Fe^{3+}]O_4^{2-}$ e.g. $NiFe_2O_4$

1.6   Magnetic interactions in ferrites

The distribution of cations over the A-sites and B-sites in the unit cell of ferrite leads to the possibility of three kinds of magnetic interactions namely, A-A, A-B and B-B sites interactions, between the nearest neighbours (Roshan et al., 2007). Instead of direct exchange interactions between the cations at A and B sites, superexchange interactions takes place in which oxygen ions mediate the interactions between the cations at A and B sites (Standley, 1962).

The inter sublattice (A-B) interactions are the strongest interactions than the intra sublattice (A-A, B-B) interactions, when the electron spins at the A and B-sites are anti-parallel and collinear. However, in the case of sufficient non-magnetic ion substitution in the ferrite, the intra sublattice (B-B) interaction becomes comparable with the inter sublattice (A-B) interaction leading to non-collinear spin structure in the B-site (Roshan et al., 2007).

In the case of collinear structure, due to the anti-parallel orientation of electron spins of the ions at A and B sites, the interaction energy is negative. The magnitude of the interaction energy depends upon the distance and angle between the ions. The favorable distances and angles for magnetic interactions between the ion pairs in spinel ferrite as reported by Gorter (Gorter, 1954) is shown in Fig.1.3. The interaction energy varies directly with angle (has its maximum when the angle is 180°) and inversely with distance.

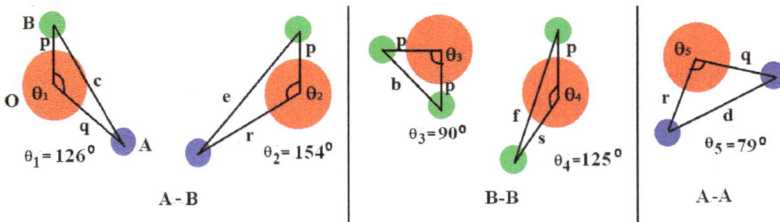

Fig.1.3 *Configuration of ion pairs in the unit cell of spinel ferrites for effective interactions between ions present at tetrahedral (A) site and at octahedral (B) site. O represents the oxygen atom.*

Based on the angles in Fig.1.3, among the three interactions, the A-B interaction has the largest magnitude. Of the two configurations for A-B interaction, θ_2 value is larger than that of θ_1 but smaller distance value of 'q' makes the first configuration as the strongest interaction. Of the two configurations for the B-B interaction, the first configuration will be effective, though the value of θ_3 is equal to 90°, since, in the second configuration the distance 's' is too large for effective interaction. Thus, overall B-B interactions are expected to be lower in strength than that of A-B interactions. The A-A interaction is the weakest, as the distance 'r' is much larger than any A-B or B-B interactions and the angle θ_5 is also much less than180°.

11

1.7 Synthesis of ferrite samples

Preparation methods play a pivotal role in the outcome of the physico-chemical properties of ferrite particles. A preparation method should have control over the desired stoichiometric ratio, particle size, morphology etc. Numerous synthesis methods are available, namely, combustion method (Nilar et al., 2013), co-precipitation (Li et al., 2010), sol-gel (Thomas et al., 2013), solid state reaction (Liu et al., 2012) and microwave sintering (Reddy et al., 2012) etc. In this research work, Mechanical alloying method followed by sintering process, co-precipitation method and molten – salt low temperature route (chemical reaction method) were adopted to synthesis the samples.

1.7.1 Mechanical alloying method

The mechanical alloying method, developed by John Benjamin in the mid-1960 belongs to the solid state powder processing technique. The mechanical alloying method is based on the concept of synthesizing materials in a non-equilibrium state by "energizing and quenching". Energizing brings the material into a highly non-equilibrium (metastable) state by some external dynamical force and quenching brings the material into a configurationally frozen state which will be used as a precursor to obtain final desired sample by subsequent sintering at high temperatures. Mechanically alloyed synthesized materials possess improved physical and mechanical characters in comparison with conventional processed materials (Suryanarayana, 2001).

1.7.2 Sintering

The samples synthesized by mechanical alloying are usually subjected to sintering process. Sintering controls the extrinsic properties such as particle size, shape, distribution, porosity, and agglomeration etc., of the samples. The sintering process is a two stage process (a) pre-sintering and (b) final sintering.

1.7.3 Pre-sintering

The as prepared samples are mixed thoroughly by grinding and then pressed into pellets using a hydraulic press followed by heat treatment for some specific duration.

1.7.4 Final sintering

The pre-sintered pellets were grounded into fine powder to reduce the particle size and to promote the mixing of any un-reacted oxides. After grinding, the powder is then pressed into required shape by applying pressure using a hydraulic press. The pellets/torroids prepared in this manner are sintered at the desired temperature for some specific duration. The final sintering increases the density and decreases the porosity of the material.

1.7.5 Co-precipitation method

Co-precipitation, a wet chemical method, possesses some peculiar advantages in synthesizing the ferrite particles. It has a very simple preparation procedure and has good control over particle size and composition. The salts that are used for the precipitation generally are nitrates, chlorides or sulphates and the alkali solution may be either ammonium hydroxide (NH_4OH) or sodium hydroxide (NaOH).

The nitrate salts in proper proportions are dissolved in doubly de-ionized water followed by stirring with the help of magnetic stirrer at an appropriate temperature. NaOH solution (as precipitant) is added to control the pH of the solution because particle size of the co-precipitated material is strongly dependent on pH of the precipitation medium and molarities of the starting precursors. The precipitations are then washed and dried to obtain the final products (Hankare et al., 2009; Nikumbh et al., 2014).

1.7.6 Molten-salt low temperature method

In the molten salt low temperature method, which is a chemical reaction method, sodium chloride plays a vital role in inhibiting the grain growth. This effective, environment friendly route is high yielding, low cost, convenient and an inexpensive method. Metal oxides obtained by grinding solid metallic salts with sodium hydroxide are different from those acquired with hydroxides in solution. The grain growth is inhibited by the presence of sodium chloride as it leads to the formation of coating surrounding the nano particle and preventing the nuclei from aggregating to large particles. The reagents are taken in a mortar and pestle in a specified molar ratio and grounded together, followed by calcination and filtering.

1.7.7 Synthesis of $Ni_{0.5}Zn_{0.5}Fe_2O_4$ nano ferrite particles by mechanical alloying

Analytical grade ferric oxide (Fe_2O_3) with purity (99.95%), nickel oxide (NiO) with purity (99.99%), and zinc oxide (ZnO) with purity (99.99%) were purchased from Alfa Aesar and were used without any purification. 16 gm of Fe_2O_3, 3.734 gm of NiO and 4.069 gm of ZnO were taken for the preparation of $Ni_{0.5}Zn_{0.5}Fe_2O_4$ nano ferrite particles by mechanical alloying via ball milling. The samples were milled for 24 hours with a speed of 1450 rpm using the SPEX 8000D shaker mill instrument in an ambient atmosphere with stainless steel ball to powder ratio of 10:1. Ten balls with 12mm diameter each were used together with 25 ml mixing load of the vial. The materials of the ball and vial were made from hardened steel. The overall reaction is

$$0.5NiO + 0.5ZnO + Fe_2O_3 \rightarrow Ni_{0.5}Zn_{0.5}Fe_2O_4 \qquad (1.1)$$

The alloyed powder was then divided equally into eight samples and each was then uniaxially pressed into pellets with 2 wt% of PVA as binder under a pressure of 7T. The samples were sintered at different sintering temperatures i.e., 700°C, 800°C, 900°C, 1000°C, 1100°C, 1200°C, 1300°C and 1400°C for 10 hours in ambient atmosphere.

1.7.8 Synthesis of $(X_xY_yZ_z)$ Fe_2O_4 ferrites samples by co-precipitation method

The $(X_xY_yZ_z)$ Fe_2O_4 ferrite samples were synthesized by co-precipitation method by substituting analytical grade nitrates for X, Y and Z with x, y and z values. The analytical grade nitrate reagents that were substituted for X, Y and Z are nickel nitrate $(Ni(NO_3)_2.6H_2O)$, zinc nitrate $(Zn(NO_3)_2.6H_2O)$, manganese nitrate $(Mn(NO_3)_2.6H_2O)$, cobalt nitrate $(Co(NO_3)_2.9H_2O)$, copper nitrate $(Cu(NO_3)_2.3H_2O)$, magnesium nitrate $(Mg(NO_3)_2.6H_2O)$ and iron nitrate $(Fe(NO_3)_3.9H_2O)$. All reagents were purchased from Alfa Aesar with purity 99.99% and were used without any purification. The samples synthesized were

$Ni_{0.8}Zn_{0.2}Fe_2O_4$ (X = Ni, x = 0.8; Y and y = 0; Z = Zn with z = 0.2).

$Mn_{0.4}Zn_{0.6}Fe_2O_4$ (X = Mn, x = 0.4; Y and y = 0; Z = Zn with z = 0.6).

$Co_{0.4}Zn_{0.6}Fe_2O_4$ (X = Co, x = 0.4; Y and y = 0; Z = Zn with z = 0.6).

$Ni_{0.53}Cu_{0.12}Zn_{0.35}Fe_2O_4$ (X = Ni, x = 0.53; Y = Cu, y = 0.12; Z = Zn, z = 0.35).

$Mg_{0.2}Cu_{0.3}Zn_{0.5}Fe_2O_4$ (X = Mg, x = 0.2; Y = Cu, y = 0.3; Z = Zn, z = 0.5).

Nitrate reagents (X, Y and Z) were taken in stoichiometric proportions according to the following equation;

$$(X_xY_yZn_z)Fe_2O_4 \tag{1.2}$$

and dissolved in 50 ml of de-ionized water. Iron nitrate with 0.5M and 10gm was used in the preparation of all samples. Aqueous sodium hydroxide (2M, 4gm) was added to the prepared solutions with constant stirring to control the pH. The precipitations were separated by centrifugation and then washed repeatedly with de-ionized water, followed by drying in an oven overnight at 60°C. The powders were finally sintered at 900°C for five hours using conventional sintering method. The molar ratio of (X+Y+Z): Fe was maintained as 1:2 for all the samples. The overall reaction to synthesize the samples by co-precipitation method is given as follows.

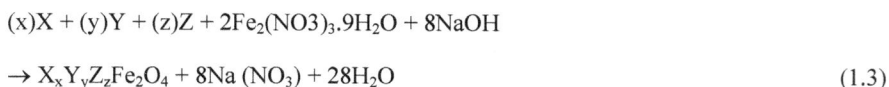

$$(x)X + (y)Y + (z)Z + 2Fe_2(NO3)_3.9H_2O + 8NaOH$$

$$\rightarrow X_xY_yZ_zFe_2O_4 + 8Na(NO_3) + 28H_2O \tag{1.3}$$

For example, the synthesis of $Ni_{0.53}Cu_{0.12}Zn_{0.35}Fe_2O_4$ nano ferrite particles is given as;

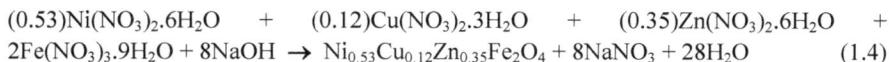

$$(0.53)Ni(NO_3)_2.6H_2O \quad + \quad (0.12)Cu(NO_3)_2.3H_2O \quad + \quad (0.35)Zn(NO_3)_2.6H_2O \quad +$$
$$2Fe(NO_3)_3.9H_2O + 8NaOH \rightarrow Ni_{0.53}Cu_{0.12}Zn_{0.35}Fe_2O_4 + 8NaNO_3 + 28H_2O \qquad (1.4)$$

The reagents, molar ratio and quantity of all reagents and pH values, are tabulated in Table 1.2.

Table 1.2 *Reagents, molar ratio and quantities of the samples*

Sample	Reagents	Molar ratio	Quantities (gm)	pH
$Ni_{0.8}Zn_{0.2}Fe_2O_4$	$Ni(NO_3)_2.6H_2O$	0.2	2.91	> 10
	$Zn(NO_3)_2.6H_2O$	0.05	0.74	
$Mn_{0.4}Zn_{0.6}Fe_2O_4$	$Mn(NO_3)_2.6H_2O$	0.1	1.25	= 9.5
	$Zn(NO_3)_2.6H_2O$	0.15	2.23	
$Co_{0.4}Zn_{0.6}Fe_2O_4$	$Co(NO_3)_2.9H_2O$	0.1	1.45	=10
	$Zn(NO_3)_2.6H_2O$	0.15	2.23	
$Ni_{0.53}Cu_{0.12}Zn_{0.35}Fe_2O_4$	$Ni(NO_3)_2.6H_2O$	0.1325	1.92	> 10
	$Cu(NO_3)_2.3H_2O$	0.03	0.36	
	$Zn(NO_3)_2.6H_2O$	0.0875	1.30	
$Mg_{0.2}Cu_{0.3}Zn_{0.5}Fe_2O_4$	$Mg(NO_3)_2.6H_2O$	0.05	0.64	= 9.5
	$Cu(NO_3)_2.3H_2O$	0.075	0.90	
	$Zn(NO_3)_2.6H_2O$	0.125	1.86	

1.7.9 Synthesis of $NiFe_2O_4$ nano ferrite particles

Nickel Ferrite nano particles were prepared by chemical reaction method. Analytical grade nickel sulphate ($NiSO_4.6H_2O$) with purity 98%, iron nitrate ($Fe(NO_3)_3.9H_2O$) with purity 99.99%, sodium hydroxide (NaOH) and sodium chloride (NaCl) were purchased from Alpha Aeser and used without any purification. The molar ratio was maintained as 1:2:8:10 for $NiSO_4.6H_2O$, $Fe(NO_3)_3.9H_2O$, NaOH and NaCl respectively. 2.628 gm nickel sulphate, 8.079 gm of iron nitrate, 3.199 gm of sodium hydroxide and 5.844 gm of sodium chloride were taken and ground together in an agate mortar and pestle for an hour. An exothermic reaction took place and color change from greenish red to brown was observed during this mixing process. NaCl was used to reduce the growth of grain into large particles (Sonali et al., 2009). The overall mechanism of the reaction is given as $Ni(SO_4).6H_2O + 8NaOH + 2Fe(NO_3)_3.9H_2O$

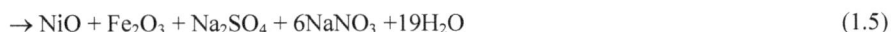

$$\rightarrow NiO + Fe_2O_3 + Na_2SO_4 + 6NaNO_3 + 19H_2O \qquad (1.5)$$

$$NiO + Fe_2O_3 \xrightarrow{700°C} NiFe_2O_4 \tag{1.6}$$

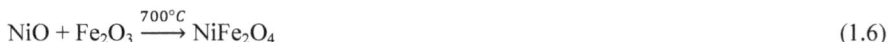

This mixture was subjected to calcination at 700°C for an hour. The powder was crushed and then washed with deionized water to remove sodium chloride and dried at 100°C for an hour to obtain the polycrystalline nickel ferrite nano particles.

1.8   Characterizations techniques

The following section will highlight the basics of various characterization techniques used in the present research work.

1.8.1   Structural studies – X-Ray diffraction

In order to obtain diffraction pattern, the essential condition is that the dimension of the scatterer and the wavelength of the incident electromagnetic wave should be compatible. To know the internal structure of the crystals, the wavelength incident on it must be comparable with the interatomic distances in the crystal.

Prof. Bragg explained the complicated process of diffraction of X-rays by a three dimensional crystal grating by supposing that reflection occurred from the various planes of atoms/molecules in the crystal. The two necessary conditions based upon which Bragg derived his law are wavelets scattered from the planes of the crystal can combine in phase only in some definite direction and not in all directions. In that particular direction, path difference between any two waves scattered from successive planes should be an integral multiple of the wavelength (Suryanarayana et al., 1998).

1.8.2   Bragg's law

To understand the Bragg's simple interpretation, consider the parallel sets of planes of the crystal separated by a distance 'd', containing the atoms repeated at regular intervals, as shown in Fig.1.4. Let 'ae' represents the incident wavefront of monochromatic X-rays of wavelength λ on the crystal at a glancing angle θ. The X-rays fall on the atoms at 'b' and 'f' and these atoms scatter the X-rays and hence, become center of disturbance and develop spherical wavelets whose envelope gives rise to the reflected wavefront. Let 'cg' represents the reflected wavefront. Now as per the first condition, the waves 'bc' and 'fg' combine and make constructive interference only in some particular direction and not in all directions. That particular direction should lie in the plane which contains the incident X-rays, which means the angle of incidence (θ) between X-rays and that crystal plane must be equal to the angle of reflection (θ) between X-rays and that crystal plane. This is the basic law for reflection of light and hence, Bragg's scattering is known as Bragg's reflection. From Fig.1.4, the path difference (p.d) between the waves 'abc' and 'efg' is

$$p.d = xf + fy \qquad (1.7)$$

where x and y are the feet of the perpendiculars drawn from b on 'ef' and 'fg' respectively.

i.e., $p.d = dsin\theta + dsin\theta = 2dsin\theta$ \qquad (1.8)

In optics, the condition for maximum intensity of reflected beam is that the path difference should be an integral multiple of the wavelength ($n\lambda$), which is the second condition in this derivation.

Thus $2dsin\theta = n\lambda$ \qquad (1.9)

This is known as Bragg's law.

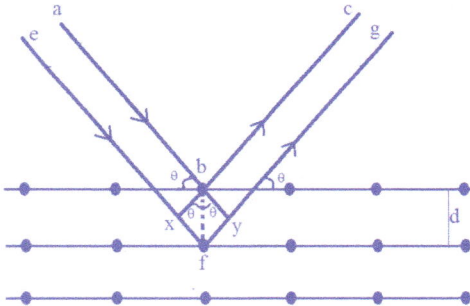

Fig.1.4 *Schematic diagram of X-ray reflection from parallel planes of a crystal.*

1.8.3 Powder diffraction

X-ray diffraction technique makes the study of crystal structure more feasible. In the powder diffraction technique, the specimen is in the form of powder rather than a single crystal. The specimen can be a solid substance of metal, ceramic, polymer, glass, thin film or a liquid, divided into very small particles. The number and size of the individual crystallites that form the specimen is important in this technique and not their degree of accretion (Georg, 2006).

One of the advantages of powder diffraction pattern is that it is representation of the bulk materials whereas in single crystal diffraction the same is not true (Dinnebier et al., 2008).

1.8.4 Thermal analysis

Thermal analysis techniques are useful to trace the changes that occur in the structural and in chemical compositions due to heat treatments of the samples. Thermogravimetric (TG) and differential thermal analysis (DTA) are the different types of thermal analysis techniques. TG is useful in analyzing the purity and the phases present in the sample, estimating the content of water and also for studying the decomposition reactions that occurred in the samples. In DTA technique, a reference material, say alumina, is also placed along with the sample, during measurement to detect any structural transformation in the sample.

1.8.5 Morphology studies - Scanning electron microscope (SEM)

The scanning electron microscope (SEM) instrument has higher magnification, larger depth of field and greater resolution than ordinary optical microscope and hence, the image of the sample produced by SEM is employed to analyze the morphology and topography of the samples under investigation.

The working principle of the SEM instrument is that a fine beam of electrons is made to fall on the samples and collisions between the electrons and atoms at or near the surface will take place and complete information about the topography and morphology of the sample can be obtained.

1.8.6 Elemental analysis and compositional studies – EDX and XRF

1.8.6.1 Energy dispersive X-ray analysis

Energy Dispersive X-Ray (EDX) analysis is employed in the elemental analysis of the sample. The interactions between the electromagnetic radiation and matter form the basis for EDX and this interaction may cause an inner shell electron to eject resulting in creation of hole. An outer shell electron may then fills this hole by making a jump to the inner shell and this process resulted in emission of an X-ray with energy corresponding to the difference between the two levels which is characteristic of that element. This allows the elemental composition of the specimen to be measured by measuring the number and energy of the X-rays emitted with the help of an energy dispersive spectrometer.

1.8.6.2 X-ray fluorescence (XRF)

X-ray fluorescence (XRF) is a non-destructive technique which is used to analyze the presence of atoms in the materials. A XRF spectrum for a single element has a number of peaks with varying intensity values. This is because the vacancy in the inner orbit, say the K shell is filled by electrons from higher energy orbits, say the L shell or M shell which in turn leads to the creation of vacancy in the L or M shell and these vacancies are filled by electrons from subsequent higher orbits. The transition of electron from higher orbits to lower orbits is governed by the laws of quantum mechanics. Thus, transition of electrons from higher orbits to the K shell is known as K transitions and similarly L transitions and M transitions etc., and hence, lead to the presence of several peaks. Among these three transitions (K, L and M series) the line that belongs to the K series possesses highest energy followed by the L and M series lines (Beckhoff et al., 2005).

1.8.7 Magnetic studies – Vibrating sample magnetometer

The vibrating sample magnetometer (VSM) is an instrument used to measure the various magnetic properties of the sample as a function of magnetic field, temperature and time. During measurements, a uniform external magnetic field (H) is applied to magnetize the given sample, followed by physical vibration of that sample by means of piezoelectric material. Due to this vibration, a magnetic flux will induce and this will generate a voltage in the sensing coil and this voltage is proportional to the magnetic moment of the sample. A plot of magnetization (M) Vs magnetic field strength (H) is generated. The software in the VSM instrument automatically extracts various parameters directly from the measured hysteresis loop.

1.8.8 Hysteresis loop

Fig.1.5 shows a typical hysteresis (M – H) loop. An increase in the magnetic field (H) causes an increase in magnetization (M) which is shown in Fig.1.5 as dotted path and further increase in magnetic field causes the magnetization to reach a maximum value 'a', known as saturation magnetization (M_s). Saturation magnetization indicates almost parallel alignment of domain inside the substance and thereafter no noticeable increase in M with any increase in H. If we withdraw the field (H) completely, then the magnetization of the substance instead of becoming zero (reaching the origin) left with some residual magnetization ('b'). To make the residual magnetization to zero, the direction of field (H) is reversed and at 'c' (Fig.1.5) the residual magnetization becomes zero. Further, increasing the field (H) in that same direction, the material is magnetized in the opposite direction and finally attains saturation at 'd' ($-M_s$). On reversing the field again and making H to zero, the substance will have residual magnetism 'e' which is numerically equal but opposite in direction to 'b'. Increasing H further, the residual ('e') magnetism is wiped out at 'f', which is equal but opposite to 'c'. From 'f', it reaches 'a' with increase in H to complete the loop (Hubert et al., 1998). If the field (H) is varied cyclically between the points 'a' and 'd', magnetization changes repeatedly along the curve 'abcdefa' but never goes to zero. Thus, if a substance is magnetized to the fullest extent, magnetization (M) does not change in phase with magnetic field (H); rather M

19

lags behind H. This lagging of M behind H is known as hysteresis and the curve 'abcdefa' is known as hysteresis loop.

Retentivity is the residual magnetization that is left in the material when the magnetic field is removed after achieving saturation. The value of B at point 'b' on the hysteresis curve as shown in the Fig.1.5 is called retentivity. If the material had not saturated magnetically then the residual magnetism and retentivity values will be different. Residual magnetism may be lower than the retentivity in that case. Coercivity is the quantity of reverse magnetic field which must be applied to a magnetic material to make the magnetic flux return to zero.

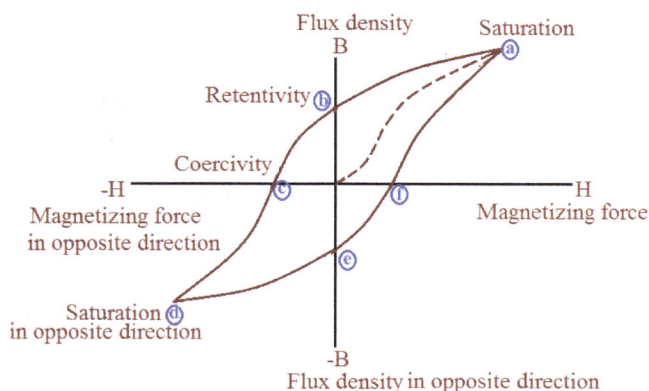

Fig.1.5 *Magnetic hysteresis loop.*

The value of H at point 'c' on the hysteresis curve as shown in the Fig.1.5 is known as coercivity. Permeability is the property of a material that describes the ease with which a magnetic flux is established in the substance (O'Handley et al., 2000).

1.8.9 Optical studies - UV-visible spectroscopy

This principle is used for quantitative measurements of optical properties of materials/compounds and it is a non-destructive technique. The wavelength range of an UV-Visible spectrometer, generally, is between 200nm (near ultra-violet) to 800nm (near infrared) and this range is very narrow when we take into account the whole electromagnetic spectrum, but still this range is very important to the researchers since, it correspond to the energy difference of the electronic states of atoms and molecules (**Perkampus** et al., 1992). When radiation, in the form of UV-Visible, interacts with the matter, a number of processes including reflection, scattering and absorbance may occur (Tony, 1996). UV-Visible spectroscopy concerns with the absorption of radiation by matter since this absorbance causes the outer electrons to move to the excited energy states from their ground states, if the energy of the incident UV radiation matches exactly with that of energy differences between two electronic states in the atom or molecule (Benfaust, 1997). An optical spectrometer records the wavelengths at which absorption occurs and presents the same in the form of graph. The absorption (A) of a solution at a particular wavelength is given by the Beer-Lambert's law

$$A = ect \qquad\qquad (1.10)$$

where e is the molar extinction coefficient characteristic of the compound at a given wavelength, c is the concentration of the compound and t is the thickness of the cell.

1.8.10   Dielectric studies – Broad band dielectric spectrometer

Broad band dielectric spectrometer (BDS) is used to analyze the electrical properties of the sample as a function of frequency and temperature. The interaction of applied electric field with the electric dipole moment and charges of the sample forms the basis for this spectroscopy.

Dielectric properties of ferrites depend mainly upon the method of preparation, composition, sintering temperature, duration of sintering, sintering atmosphere and grain size of the sample (Rangamohan et al., 1999; Ravinder et al., 2000). The study of dielectric materials reveals how an electric field conducts inside that material through polarization. Dielectrics are characterized by dielectric constant, dielectric loss, dielectric strength and resistivity.

1.8.11   Polarization

The applied electric field has influence over the electrical behavior of a dielectric material by exerting a force on the positively charged nucleus and negatively charged electrons of an atom in dielectric material. Polarization in dielectric is defined as a process of producing electrical dipoles inside the dielectric material by applying an external electric field.

The polarizations in dielectric material occur due to different microscopic mechanisms and they are (i) electronic polarization (ii) ionic polarization (iii) orientation or dipolar polarization and (iv) space - charge or interfacial polarization (Navneet et al., 2011).

1.8.12 Effects of frequency on polarization

When an alternating electric field is applied across a dielectric material, polarization occurs as a function of time.

Electronic polarization is the fastest polarization which will complete at the instant the field is applied since, the electron being lighter elementary particle, responds to the applied electric field instantaneously than ions that present in the dielectric material. It occurs at all frequencies including optical range ($\approx 10^{15}$ Hz).

Ion is heavier than electron and hence, ionic polarization is little slower than electronic polarization. In addition, the frequency of applied electrical field with which the ions will be displaced is equal to frequency of lattice vibration ($\approx 10^{13}$ Hz). Ionic polarization does not occur at optical frequency since the time required for lattice vibrations is nearly 100 times more than the period of applied field at optical frequency.

Orientational polarization is slower than the ionic and occurs only at electrical frequencies (audio and radio frequency, $\approx 10^6$ Hz) which are smaller than the infrared frequencies.

Space-charge polarization is the slowest because ions have to diffuse over several atomic distances and occurs only at power frequencies ($\approx 10^2$ Hz).

Thus, at low frequencies, the total polarization value is very high whereas at optical frequencies the total polarization value is low.

1.8.13 Dielectric constant

If the air medium between the two plates of a condenser is replaced by a dielectric medium then capacitance of the condenser increases. The dielectric constant or relative permittivity (ε') is defined as

$$\varepsilon' = C_m/C_a \tag{1.11}$$

where C_m and C_a are respectively capacitance with given dielectric medium and in vacuum (Raghavan, 2012). According to International System of unit (SI), ε' is defined as

$$\varepsilon' = \varepsilon/\varepsilon_o \tag{1.12}$$

where ε and ε_o are the absolute permittivity and permittivity of free space respectively. Dielectric constant depends upon the arrangement of electrons in the atom. If the dielectric medium has high dielectric constant, then the dielectric medium will be easily polarized and behaves as a good electrical insulator.

1.8.14 Frequency dependence of dielectric constant

The frequency of direct current is zero, whereas the frequency of alternating current has some definite value thus, the direction of polarization changes with time when the dielectric material is subjected to an alternating electric field. Each type of electric polarization has its own minimum reorientation time which depends on the ease with which the particular dipoles are capable of realignment. Relaxation frequency is inversely proportional to the minimum reorientation time and a dipole cannot follow the frequency of the alternating current, if the frequency of alternating current is much higher than its relaxation frequency and will not make a contribution to the dielectric constant (Abdullah et al., 2010).

1.8.15 Dielectric strength

Dielectric strength refers to the maximum electric stress that the dielectric material can withstand without losing its insulating properties (Raghavan, 2012). At breakdown, the applied voltage is sufficient to liberate the bound electrons and if the applied voltage is high enough to accelerate the liberated electrons then avalanche breakdown occurs. Dielectric strength is directly proportional to the thickness of the material and inversely proportional to the temperature, humidity and applied frequency. Dielectric loss is the amount of the electrical energy that is dissipated in the form of heat energy and dielectric loss of a dielectric material is the least when the dielectric material is subjected to a direct current field (Kakani et al., 2004)

1.9 Characterizations of ferrites

The ferrites synthesized in the present research work were characterized for

1. *Structural and average particle size analysis using X-ray diffraction studies.*

2. *Thermal analysis using TG/DTA studies.*

3. *Morphological analysis using scanning electron microscope (SEM) studies.*

4. *Elemental analysis using energy dispersive X-ray (EDX) studies.*

5. *Compositional analysis using X-ray fluorescence (XRF) studies.*

6. *Magnetic properties analysis using vibrating sample magnetometer (VSM) studies.*

7. *Dielectric properties using broad band dielectric (BDS) spectrometer studies.*

8. *Optical properties using UV-Vis characterization studies.*

Room temperature raw X-Ray diffraction of all powder ferrite samples were recorded at the Sophisticated Test and Instrumentation Centre (STIC), Cochin University of Science and Technology (CUSAT), Cochin, India, using Bruker AXS D8 advance X-ray diffractometer with Cu radiation (λ=1.5406 Å). The samples were exposed to the radiation with a primary beam power of 40 kV and 35 mA with a step scan of 0.02° in 2θ and 2θ range from 25° to 120°.

Thermogravimetric studies were carried out on samples $Ni_{0.5}Zn_{0.5}Fe_2O_4$, sintered from 700 - 1000°C and on the sample $Ni_{0.8}Zn_{0.2}Fe_2O_4$ sintered at 900°C. The TG/DTA analysis were studied in the temperature range of 40–730°C at the rate of 5°C/min in static air atmosphere using thermal analyzer (Model: Perkin Elmer, Diamond TG/DTA) at STIC, CUSAT, Cochin, India.

The morphological analysis of the powder samples were studied with a scanning electron microscope (SEM) (Hitachi, Model: S-3400 N) at Central Instrumentation Facility (CIF), Pondicherry University, India.

Energy Dispersive X-Ray (EDX) analyses of all pelletized samples, were recorded at Sophisticated Analytical Instruments Facility (SAIF), IIT-Madras, India, using Model: Quanta 200 FEG.

The X-ray fluorescence (XRF) characterizations of all powder samples were carried out at CIF, Pondicherry University, India, using Bruker, Model: S4 pioneer, for the estimation of stoichiometric proportions and the elemental composition of samples concerned.

Room temperature magnetic properties of all powder samples were recorded at SAIF, IIT-Madras, India, using VSM (Model: LakeshoreVSM 7410).

The broadband dielectric spectrometer (BDS) characterization was carried out on $Ni_{0.5}Zn_{0.5}Fe_2O_4$ samples sintered from 700 to 1200°C and $NiFe_2O_4$ sample at CIF, Pondicherry University, Puducherry, India, using NOVOCONTROL Technologies GmbH & Co., Germany, Model: Concept 80. The dielectric constant (both real and imaginary part), dielectric loss factor and ac conductivity, at room temperature, were studied in the frequency range from 0.1Hz to 15MHz.

The UV-Vis characterization of all samples under investigation in this research work, in order to evaluate the optical band gap energy, was carried out by using UV-Vis-NIR Spectrophotometer (Model: Varian, Cary 5000) at STIC, CUSAT, Cochin, India. The range of wavelength employed was 200 – 800nm.

1.10 Methodologies employed in this research work

This section deals with the methodologies employed, namely Rietveld method (Rietveld, 1969) and MEM (Gull and Daniel, 1978) in this research work, respectively to refine the raw X-ray diffraction data and to determine the bond strength between the A-site and B-site in the unit cell of ferrites.

1.10.1 Rietveld method

Rietveld method (Rietveld, 1969) was devised by Hugo Rietveld (Rietveld, 1969). This method reduces the impact of overlapped and degenerate peaks by taking into account of the entire powder pattern for crystal structure refinement. This powder profile fitting method refines the background, pseudo-voigt, asymmetry, preferred orientation, scale and lattice parameters using a least-squares approach, until the agreement between the calculated and measured diffraction profiles are optimized (Saravanan et al., 2012; Saravanakumar et al., 2012; Saravanan et al., 2014).

1.10.2 Theory of Rietveld method

Rietveld method (Rietveld, 1969) requires high quality powder diffraction data of the samples in step scan mode possibly with 0.02° increment in 2θ range from 20° to 120°. The principle of Rietveld method (Rietveld, 1969) is to minimize a function

$$\chi = \sum_i w_i \{y_i^o - \frac{1}{c} y_i^c\}^2 \qquad (1.13)$$

where the terms w_i and c refers respectively the statistical weight and overall scale factor such that

$$y^c = cy^o. \qquad (1.14)$$

where y_i^o and y_i^c respectively the observed and calculated intensity at a certain 2θ angle. Bragg's reflection contribute to the observed intensity y_i^o whereas the calculated intensity y_i^c is evaluated using the formula (Saravanan et al., 2012)

$$y_i^c = S\sum_k L_k |F_k|^2 \varphi(2\theta_i - 2\theta_k)p_k A + y_{bi} \qquad (1.15)$$

where S – scale factor

k – Miller indices (hkl) of a Bragg reflection

L_k – Lorentz, polarization and multiplicity factors

F_k – structure factor

$\Phi (2\theta_i - 2\theta_k)$ – reflection profile function

p_k – preferred orientation function

A – absorption factor

y_{bi} – background intensity

The systemic and accidental peak overlap is overcome in the Rietveld method (Rietveld, 1969) by minimizing the weighted sum of the squared differences between the observed and the calculated pattern using least square methods.

1.10.3 Rietveld refinement

Software JANA 2006 (Petříček et al., 2006) has been employed to perform the Rietveld refinement (Rietveld, 1969) in this present research work. JANA 2006 (Petříček et al., 2006) is a crystallographic program for structural analysis of crystals which are periodic in three or more dimensions. This software calculates structures having up to three modulation vectors from the diffraction data. Structure solution is achieved by using either built in charge flipping algorithm or external direct methods program. JANA 2006 (Petříček et al., 2006) can refine the structures having more than one phase. During the refinement, the observed profiles are matched with the pseudo-voigt (Wertheim, 1974) profiles and the profile asymmetry is introduced by employing multi term Simpson rule integration devised by Howard (Howard, 1982) that incorporates symmetric profile shape function with different coefficient for weights and peak shift. Corrections for preferred orientation is employed using March – Dollase function (March, 1932; Dollase, 1986). Finally, the structure factors evolved from the Rietveld refinements were further utilized for the estimation of charge density in the unit cell.

Analysis of powder diffraction pattern requires collection of high quality powder diffraction data followed by indexing the powder diffraction pattern and intensity determination (Georg, 2006). Maximum entropy method (MEM) combined with Rietveld method (Rietveld, 1969) is a potential tool for the analysis of powder diffraction data. The MEM/Rietveld (Takata, 2008) method compliments both the Rietveld analysis (Rietveld, 1969) of the observed powder diffraction profile and the MEM charge density imaging (Saravanakumar et al., 2012).

1.10.4 Maximum entropy method (MEM)

The knowledge of physical and chemical properties of any material is a key factor in understanding that material. In a molecule, all atoms are held together by means of bonds between the valence electrons. Hence, a clear understanding of bonding between atoms in a molecule is essential and that understanding could be enhanced if we visualize those bonds. Fourier method is one such tool for visualization of electron bonding system but it has two main drawbacks (i) it requires a structural model which should be as close as to the real crystal system and (ii) suffers from series termination error. MEM overcomes these difficulties. Mid-bond electron density values can be evaluated from the electron density distribution and this value (indicating the strength of the material) is a key parameter in evaluating a material for various applications.

Claude Shannon, a radar official in the government of United States of America, showed keen interest in efficient way of encoding and decoding a message. Later Gull and Daniel developed a method known as Maximum Entropy Method (MEM) in the field of radio astronomy to enhance the information obtained from noisy data (Gull and Daniel, 1978). Now, it finds applications in all type of spectroscopic or diffraction data and known as "imaging of diffraction data" (Sakata, 1996). De Vries (De Vries et al., 1994) and Gilmore (Gilmore, 1996) reported that MEM effectively deemphasizes the series termination error, a main drawback of Fourier synthesis method and MEM is free from structural model. A high quality data set with a limited number of reflections can give reasonable MEM maps but the success of MEM depends purely on the quality of the experimental information and any bias will reflect in the probable solutions (Israel et al., 2004).

1.10.5 Implication of MEM

The solution of MEM has the following important implications in the field of crystallography.

- MEM, along with the Rietveld technique (Rietveld, 1969), is used to locate the atom in the unit cell in the case of an unknown structure.

- Helps to determine the electron charge density in unknown as well as in known structures (Sandervan et al., 2009).

- MEM analysis not only gives the electron density distribution but also the accurate bond-length distribution, the radii of atoms/ions, etc. (Saravanan, 2009).

- MEM electron densities are always positive, and even with limited number of data one can determine reliable electron densities (Saravanan et al., 2007).

1.10.6 Principle of maximum entropy method (MEM)

The principle of maximum entropy method (MEM) states that the distribution which maximize the information entropy corresponds to that which is statistically most likely to occur and the distribution ensures that the largest remaining uncertainty should be consistent with the constraints.

1.10.7 Theory of MEM

The maximum entropy method was first introduced into crystallography by Collins (Collins 1982). Maximum entropy method (MEM), an approach based on statistics, is used to construct the electron density map, from a set of X-ray structure factors, between atoms or ligands.

Consider discretization of any trail density on a grid over a unit cell, in which the electron density is sampled at the points of a $N_1 \times N_2 \times N_3 = N_{pix}$.

The entropy of the trial density is defined as

$$S = -\sum \rho'(r) ln\left(\frac{\rho'(r)}{\tau'(r)}\right) \qquad (1.16)$$

where the probability $\rho'(r)$ and prior probability $\tau'(r)$ of trial density are related to the actual electron density in a unit cell as

$$\rho'(r) = \frac{\rho(r)}{\sum_r \rho(r)} \text{ and } \tau'(r) = \frac{\tau(r)}{\sum_r \tau(r)} \qquad (1.17)$$

where $\rho(r)$ and $\tau(r)$ are the electron density and prior electron density at a fixed r in a unit cell respectively. In the present theory, the actual densities are treated hereafter instead of normalized densities, and $\rho'(r)$ become $\tau'(r)$ when there is no information. The normalization condition for $\rho'(r)$ and $\tau'(r)$ is

$$\sum \rho'(r) = 1 \text{ and } \sum \tau'(r) = 1 \qquad (1.18)$$

The entropy S is maximized subject to different constraint

$$C = \frac{1}{N_F} \sum_{i=1}^{N_F} \frac{|F_{obs}(H) - F_{MEM}(H)|^2}{\sigma^2(H)} \qquad (1.19)$$

where N_F is the total number of reflections used in the analysis of MEM. $F_{obs}(H)$ are experimental structure factors with standard deviations σ_H and $F_{MEM}(H)$ are structure factors corresponding to the final calculated distribution Q of the electron density. The entropy is maximized by subjecting the constraint function C = 1 by using the method of undetermined Lagrange multiplier (λ).

Hence,

$$Q = S - \lambda C \qquad (1.20)$$

$$Q = -\sum \rho'(r) ln\left(\frac{\rho'(r)}{\tau'(r)}\right) - \frac{\lambda}{N_F} \sum_{i=1}^{N_F} \frac{|F_{obs}(H) - F_{MEM}(H)|^2}{\sigma^2(H)} \qquad (1.21)$$

When (dQ/dρ) = 0 and substituting the approximation ln x = x − 1 we get,

$$\rho(r_i) = \tau(r_i) \exp\left[\left(\frac{\lambda F_{000}}{N}\right)\left(\sum \frac{1}{\sigma(H)^2}\right)|F_{MEM}(H) - F_{obs}(H)| \exp(-2\pi j H.r)\right] \qquad (1.22)$$

where F_{000} represents the total number of electrons (Z) in the unit cell. The above equation is solved by substituting

$$F_{MEM}(k) = v \Sigma \tau(r) \exp(-2\pi i H.r) dV \tag{1.23}$$

This approximation can be called zeroth order single pixel approximation (ZSPA). By using this approximation the right-hand side of Eq.(1.23) becomes independent of $\tau(r)$ and Eq.(1.23) can be solved in an iterative way starting from a given initial density for the prior distribution.

1.11 Theories involved in the present research work

Theoretical concepts involved in the present research work, in estimating the cation distribution between the A-site and B-site in the unit cell of ferrite along with various structural parameters, magnetic, dielectric and optical properties are described here.

1.11.1 Estimation of cation distribution

The exact distribution of cation over the available A-site and B-site in the prepared nano ferrite is determined using XRD data and by comparing the observed X-ray intensities from specific Bragg planes with the calculated intensities evaluated using Buerger formula (Buerger, 1960)

$$I_{hkl} = |F_{hkl}| \, PL_p \tag{1.24}$$

where I_{hkl} is the calculated intensity of the hkl plane, F_{hkl} is the structure factor, P is the multiplicity factor and L_p is the Lorentz-polarization factor of the corresponding hkl plane. The multiplicity factor and Lorentz polarization factor values can be taken from the literature (Cullity, 1959). The structure factors of hkl planes (220), (440), (400) and (422) are used in arriving at the cation distribution, since (220) and (440) are sensitive to cation distribution at the A-site while (400) and (422) are sensitive to cation distribution at the B-site (Patange et al., 2012). Hence, the ratios of observed and calculated intensities viz., I_{220}/I_{400}, I_{220}/I_{440}, I_{422}/I_{440} and I_{400}/I_{440} are considered in this work. Each ion has some preferential site to occupy in the unit cell of ferrite when it is in bulk form but in the nanometer regime the site preference is not strictly adhered. Hence, in order to determine the cation distribution one has to evaluate the intensity ratios of above said hkl planes for all possible combination of ions in both A-site and B-site. Thus, we assume a general equation as given below to arrive at the cation distribution

$$(X_\alpha^{2+} Y_\beta^{2+} Z_\gamma^{2+} Fe_{(1-\alpha-\beta-\gamma)}^{3+})^A \, [X_\delta^{2+} Y_\sigma^{2+} Z_\omega^{2+} Fe_{(1+\alpha+\beta+\gamma)}^{3+}]^B O_4 \tag{1.25}$$

29

where X, Y and Z are any substituted divalent ions, and α, β and γ are the amount of cations occupying at A-sites for X, Y and Z atoms respectively and δ, σ and ω represents the same but at B-sites for X, Y and Z atoms respectively with the restriction that $(\alpha+\delta+\beta+\sigma+\gamma+\omega = 1)$. The structure factor equations for the above reflections are given below (Patange et al., 2012)

$$F_{220} = 8f_a + 16f_o(1+ \cos2\pi4u) \tag{1.26}$$

$$F_{400} = 8f_a - 16f_b - 32f_o\cos(8\pi u) \tag{1.27}$$

$$F_{422} = 8f_a + 32f_o\cos(8\pi u)\cos^2(4\pi u) \tag{1.28}$$

$$F_{440} = 8f_a + 16f_b + 32f_o\cos^2(8\pi u) \tag{1.29}$$

where f_a is the scattering factor of tetrahedral cations, f_b is the scattering factor of octahedral cations, f_o is the scattering factor of oxygen atom and u is the oxygen positional parameter. The scattering factor for the tetrahedral and octahedral site is based on the assumed cation distribution (Eq.1.25) and given as

$$f_a = \alpha f_X + \beta f_Y + \gamma f_Z + (1-\alpha-\beta-\gamma)f_{Fe} \tag{1.30}$$

and

$$f_b = \delta f_X + \sigma f_Y + \omega f_Z + (1+\alpha+\beta+\gamma)f_{Fe} \tag{1.31}$$

The atomic scattering factor for nickel, zinc, iron and oxygen atoms were taken from the international tables for X-ray crystallography (MacGillavry et al., 1968). From Eq.(1.24) the calculated intensity can be obtained. The X-ray intensity ratios of I_{220}/I_{400}, I_{220}/I_{440}, I_{422}/I_{440} and I_{400}/I_{440} were found out for various cation distribution combinations and those matching closely with observed intensity ratios are considered as cation distributions in the sample and using this cation distribution raw XRD data is refined by employing the Rietveld refinement method (Rietveld, 1969) using the JANA 2006 software (Petřiček et al., 2000).

The radius of the ion at the A-site and B-site can be calculated from both XRD data and distribution of cations. The tetrahedral site radii ($r_{A(XRD)}$) and octahedral site radii ($r_{B(XRD)}$) from XRD data, using experimental the lattice parameter (a_{exp}) and oxygen positional parameter (u), (Shinde et al., 2014)

$$r_{A(XRD)} = a_{exp}(\sqrt{3})(u - 0.25) - R_o. \tag{1.32}$$

$$r_{B(XRD)} = a_{exp}(\tfrac{5}{8} - u) - Ro. \tag{1.33}$$

Shared edge length d_{AES} and d_{BES} at A and B-sites and unshared edge length d_{BEU} at the B-site are calculated using (Attia, 2006)

$$d_{AES} = a_{exp}(\sqrt{2})(2u - 0.5) \tag{1.34}$$

$$d_{BES} = a_{exp}(\sqrt{2})(1 - 2u) \tag{1.35}$$

$$d_{BEU} = a_{exp}[4u^2 - 3u + (11/16)^{1/2}] \tag{1.36}$$

The volume ($V = a_{exp}^3$) of the unit cell and the X-ray density (D_x) of all the prepared samples were evaluated respectively using the relations (Nikam et al., 2015)

$$D_x = \frac{ZM}{NV} \tag{1.37}$$

where Z (= 8) is the number of molecules in the unit cell, M is the molecular weight, N is the Avogadro's number (6.023×10^{23} mol^{-1}) and V is the volume of the unit cell. Average grain size (t) is calculated using the Scherrer formula (Nikam et al., 2015)

$$t = \frac{0.9\lambda}{\beta cos\theta} \tag{1.38}$$

where λ is the X-ray wavelength (Cu Kα radiation = 1.54 Å), β is the line broadening at half the maximum intensity in radians and θ is the Bragg's angle in degrees. $r_{A(CD)}$ and $r_{B(CD)}$ values can also be estimated using the relation (Attia, 2006) based on cation distribution (Eq.1.25) in the system

$$r_{A(CD)} = \alpha r_X + \beta r_Y + \gamma r_Z + (1-\alpha-\beta-\gamma)r_{Fe} \tag{1.39}$$

$$r_{B(CD)} = 0.5 \times [\delta r_X + \sigma r_Y + \omega r_Z + (1+\alpha+\beta+\gamma)r_{Fe}] \tag{1.40}$$

The theoretical lattice parameter (a_{the}) is calculated using (Kurmude et al., 2014)

$$a_{the} = \frac{8}{3\sqrt{3}}(r_A + R_o) + \sqrt{3}\,(r_B + R_o) \tag{1.41}$$

where R_o (1.32Å) (Azaroff, 2009) is the radius of oxygen ion, $r_{A(CD)}$ and $r_{B(CD)}$ are the radius of the A and B site ions obtained theoretically from the knowledge of cation distribution.

The interionic distances between the cation and anion ($Me^{2+}- O$)

$$p = a_{exp} \left(\frac{1}{2} - u\right) \tag{1.42}$$

$$q = a_{exp} \left(u - \frac{1}{8}\right)\sqrt{3} \tag{1.43}$$

$$r = a_{exp}\left(u - \frac{1}{8}\right)\sqrt{11} \tag{1.44}$$

$$s = \frac{a_{exp}}{3}\left(u + \frac{1}{2}\right)\sqrt{3} \tag{1.45}$$

and between the cations ($Me^{2+} - Me^{2+}$)

$$b = \frac{a_{exp}}{4}\sqrt{2} \tag{1.46}$$

$$c = \frac{a_{exp}}{8}\sqrt{11} \tag{1.47}$$

$$d = \frac{a_{exp}}{4}\sqrt{3} \tag{1.48}$$

$$e = \frac{3a_{exp}}{8}\sqrt{3} \tag{1.49}$$

$$f = \frac{a_{exp}}{4}\sqrt{6} \tag{1.50}$$

as shown in Fig.1.3, are calculated using the experimental value of lattice constant (a_{exp}) and oxygen positional parameter (u) using the Eq.(1.42 – 1.50) (Lakhani et al., 2011).

The bond angles (Fig.1.3) are given by Eq.(1.51 – 1.55) (Lakhani et al., 2011)

$$\theta_1 = cos^{-1}\left[\frac{p^2 + q^2 - c^2}{2pq}\right] \tag{1.51}$$

$$\theta_2 = cos^{-1}\left[\frac{p^2 + r^2 - e^2}{2pr}\right] \tag{1.52}$$

$$\theta_3 = cos^{-1}\left[\frac{2p^2 - b^2}{2p^2}\right] \tag{1.53}$$

$$\theta_4 = cos^{-1}\left[\frac{p^2 + s - f^2}{2ps}\right] \tag{1.54}$$

$$\text{and } \theta_5 = cos^{-1}\left[\frac{r^2 + q^2 - d^2}{2rq}\right] \tag{1.55}$$

1.11.2 Estimation of magnetic properties of ferrites

Various magnetic properties such as saturation magnetization (M_s), coercivity (H_c), retentivity (M_r), permeability (μ), squareness ratio (M_r/M_s) were evaluated from the hysteresis curve. The critical size of the domains can be calculated using the relation (Gao et al., 2010)

$$D_{cr} = 9\varepsilon_p/2M_s^2 \text{ where } \varepsilon_p = (2k_BT_C|K_1|/a)^{1/2} \tag{1.56}$$

is known as surface energy of the domain wall, M_s is saturation magnetization, k_B is the Boltzmann constant, T_C is the Curie temperature, $|K_1|$ is the absolute value of magnetocrystalline anisotropy constant and a is the lattice parameter.

The magnetic moment per formula unit in Bohr magneton (μ_B) is calculated using the following relation from the hysteresis curve (Choodamani et al., 2013)

$$\mu_B^H = \left\{ \frac{molecular\ weight}{5585} \right\} \times M_s \qquad (1.57)$$

The magnetic moment per formula unit in Bohr magneton (μ_B) can also be estimated from the cation distribution, known as Neel's magnetic moment (μ_B^N) using the following relation (Choodamani et al., 2013)

$$\mu_B^N = (M_B - M_A) \qquad (1.58)$$

where M_B and M_A are sublattice magnetizations. The disagreement between μ_B^H and μ_B^N indicates the existence of canted spin arrangement in the octahedral (B) sublattice, known as Yaffet-Kittel (YK) angle, which is calculated using the relation (Choodamani et al., 2013)

$$\mu_B^H = M_B (\cos \alpha_{YK} - M_A) \qquad (1.59)$$

1.11.3   Estimation of optical band gap energy

UV-vis absorption analysis is one of the most frequently used methods for optical characterization of the material under investigation. The optical band gap energy of the samples is determined by using the Tauc relation (Tauc et al., 1966; Pancove, 2010)

$$(\alpha h\upsilon) \sim (h\upsilon - E_g)^n \qquad (1.60)$$

where α is the absorption coefficient, $h\upsilon$ is the energy of the incident photon, E_g is the band gap energy of the respective samples and n is a constant whose value depends on the type of transition of electron. In the present study the value of $n = \frac{1}{2}$. The value of the band gap energy is calculated from the linear interpolation of the photon energy against $(\alpha h\upsilon)^2$.

1.11.3.1 Theory

The optical band gap can be determined from the absorption spectrum using the method proposed by Wood and Tauc (Wood and Tauc, 1972). The optical band gap energy is associated with the absorbance (α) and photon energy (hv) by the following relation,

$$\alpha h v = A(h v - E_g)^n \qquad (1.61)$$

where α is the absorbance, hv is photon energy, E_g is the optical band gap, A is a constant and 'n' is an index which takes different values for different transition types of transitions, n = 1/2 for direct allowed transition, n = 3/2 for direct forbidden transition, n = 2 for indirect allowed transition and n = 3 for indirect forbidden transition, responsible for the reflection. For direct band gap materials, Eq.(1.61) can be written as,

$$\alpha hv = A(hv - E_g)^{1/2} \qquad (1.62)$$

By squaring the Eq.(1.62),

$$(\alpha hv)^2 = A(hv - E_g) \qquad (1.63)$$

$$(\alpha E)^2 = Ahv - AE_g \qquad (1.64)$$

The above equation resembles the equation of a straight line,

$$y = mx + C \qquad (1.65)$$

Comparing Eqs.(1.64) and (1.65) then

$$y = (\alpha E)^2 \qquad (1.66)$$

If y=0, then the Eq.(1.65) becomes,

$$mx + C = 0 \qquad (1.67)$$

$$A(hv - E_g) = 0 \qquad (1.68)$$

In the above equation the constant A cannot be equal to zero, then,

$$A(hv - E_g) = 0 \qquad (1.69)$$

$$E_g = hv \qquad (1.70)$$

Eq.(1.70) gives the optical band gap energy, which can be evaluated using the Tauc's procedure, by fitting a line through the linear portion of the band edge transition. Using UV-vis absorption data, a graph can be drawn by taking energy (hv) in x-axis, and $(\alpha hv)^2$ in y-axis. From this Tauc plot, the extrapolation of the tangent of linear portion of the curve to x-axis, will give the value of optical band gap E_g.

1.11.4 Estimation of dielectric properties

The dielectric constant (both real and imaginary part), dielectric loss factor and ac conductivity, at room temperature, were evaluated using the relation (Nasir et al., 2011)

$$\varepsilon' = Cd/\varepsilon_o A \tag{1.71}$$

where C is the capacitance of the pellet, d is the thickness and A is the area of cross section of the pellet. ε_o is the permittivity of the free space. Dielectric loss factor (tanδ) of the samples were measured using the relation (Nasir et al., 2011)

$$\tan\delta = \varepsilon''/\varepsilon' \tag{1.72}$$

where ε'' is the imaginary part of dielectric constant. The ac conductivity (σ_{ac}) of the samples were determined using the relation (Nasir et al., 2011)

$$\sigma_{ac} = \omega\varepsilon'\varepsilon_o \tan\delta. \tag{1.73}$$

1.12 Photographs of samples synthesized in this research study

The $Ni_{0.5}Zn_{0.5}Fe_2O_4$ samples synthesized in the present study, using mechanical alloying method are shown in Fig.1.6 and the samples namely $Mg_{0.2}Cu_{0.3}Zn_{0.5}Fe_2O_4$, $Mn_{0.4}Zn_{0.6}Fe_2O_4$, $Ni_{0.53}Cu_{0.12}Zn_{0.53}Fe_2O_4$, $Co_{0.4}Zn_{0.6}Fe_2O_4$ and $Ni_{0.8}Zn_{0.2}Fe_2O_4$ synthesized using co-precipitation method and $NiFe_2O_4$ sample molten-salt chemical route are shown in Fig.1.7.

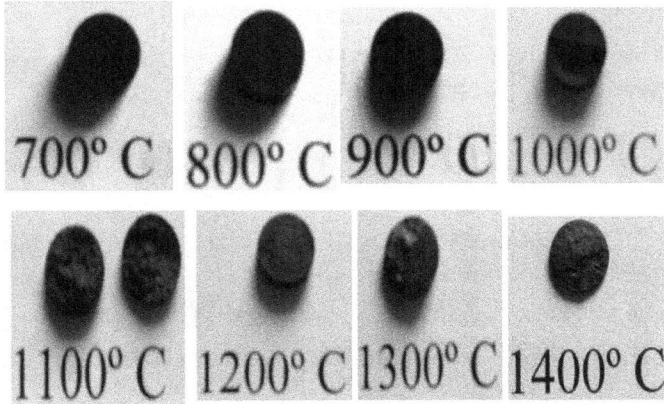

Fig.1.6 *Photographs of mechanically alloyed $Ni_{0.5}Zn_{0.5}Fe_2O_4$ samples sintered from 700 - 1400°C.*

$Mg_{0.2}Cu_{0.3}Zn_{0.5}Fe_2O_4$ $Mn_{0.4}Zn_{0.6}Fe_2O_4$ $NiFe_2O_4$

$Ni_{0.53}Cu_{0.12}Zn_{0.35}Fe_2O_4$ $Co_{0.4}Zn_{0.6}Fe_2O_4$ $Ni_{0.8}Zn_{0.2}Fe_2O_4$

Fig.1.7 *Images of samples synthesized by co-precipitation method ($Mg_{0.2}Cu_{0.3}Zn_{0.5}Fe_2O_4$, $Mn_{0.4}Zn_{0.6}Fe_2O_4$, $Ni_{0.53}Cu_{0.12}Zn_{0.53}Fe_2O_4$, $Co_{0.4}Zn_{0.6}Fe_2O_4$ and $Ni_{0.8}Zn_{0.2}Fe_2O_4$) and by chemical reaction method ($NiFe_2O_4$).*

References

[1] Abdullah DM, Khalid MB, Vivek V, Siddiqui WA, Kotnala RK (2010) J Alloys Compd 493:553-560. https://doi.org/10.1016/j.jallcom.2009.12.154

[2] Attia SM (2006) Egypt J. Solids 29:329-339.

[3] Azaroff LV (2009) Introduction to solids TMH Edition, Tata McGraw Hill Education Private Limited New Delhi.

[4] Beckhoff B, Kanngießer B, Langhoff N, Wedell R, Wolff H (Eds.) (2005) Handbook of Practical X-Ray Fluorescence Analysis, Springer.

[5] Benfaust (1997) Modern Chemical Techniques: An Essential Reference for Students and Teachers, Unilever, Royal society of Chemistry.

[6] Buerger MJ (1960) Crystal Structure Analysis Wiley New York.

[7] Candeia RA, Souza MAF, Bernardi MIB, Maestrelli SC, Santos IMG, Souza AG, Longo E (2006) Mater Res Bull 41:183-190. https://doi.org/10.1016/j.materresbull.2005.07.019

[8]   Choodamani C, Nagabhushana GP, Ashoka S, Prasad BD, Rudraswamy B, Chandrappa GT (2013) J Alloy Compd 578:103-109. https://doi.org/10.1016/j.jallcom.2013.04.152

[9]   Collins DM Nature 298:49-51.

[10]   Cullity BD (1959) Elements of X-Ray Diffraction Addision-Wesley USA.

[11]   Daliya SM, Ruey-Shin J (2007) Chem Eng J 129: 51-65. https://doi.org/10.1016/j.cej.2006.11.001

[12]   De Vries RY, Briels WJ, Feil D (1994) Acta Crystallogr A 50:383-391. https://doi.org/10.1107/S0108767393012802

[13]   Dinnebier RE, Billinge SJL (Eds) (2008) Powder Diffraction: Theory and Practice RSC publishing.

[14]   Dollase WAJ (1986) Appl Crystallogr 19:267-272. https://doi.org/10.1107/S0021889886089458

[15]   Gao P, Rebrov EV, Verhoeven TMWGM, Schouten JC, Kleismit R, Kozlowski G, Cetnar J, Turgut Z, Subramanyam G (2010) J Appl Phys 107 044317-1-8.

[16]   Georg Will (2006) Powder Diffraction Springer-Verlag Berlin.

[17]   Gilmore CJ (1996) Acta Crystallogr A 52:561-589. https://doi.org/10.1107/S0108767396001560

[18]   Gorter EW (1954) Philips Res Rept 9:295-321.

[19]   Gull SF, Daniel GJ (1978) Nature 272:686-690. https://doi.org/10.1038/272686a0

[20]   Hankare PP, Jadhav SD, Sankpal UB, Patil RP, Sasikala R, Mulla IS (2009) J Alloys Comp 488(1):270-272. https://doi.org/10.1016/j.jallcom.2009.08.103

[21]   Hemeda OM, Mostafa NY, AbdElkader OH, Ahmed MA (2014) J Magn Magn Mater 364:39-46. https://doi.org/10.1016/j.jmmm.2014.03.061

[22]   Howard CJ (1982) J Appl Crystallogr 15:615-620. https://doi.org/10.1107/S0021889882012783

[23]   Hubert A, Schäfer R (1998) Magnetic Domains Springer-Verlag Berlin.

[24]   Israel S, Saravanan R, Rajaram RK (2004) Physica B 349:390-400. https://doi.org/10.1016/j.physb.2004.04.068

[25]   Izumi F, Dilanian RA (2002) Recent research developments in Physics part II Vol.3 Transworld Research Network, Trivandrum.

[26] Jiles C.David (1998) Introduction to Magnetism and Magnetic Materials 2nd edition, Chapman & Hall/CRC.

[27] Kakani SL, Kakani A (2004) Material Science New Age International publishers.

[28] Kurmude DV, Barkule RS, Raut AV, Shengule DR, Jadhav KM (2014) J Supercond Nov Magn 27(2):547-553. https://doi.org/10.1007/s10948-013-2305-2

[29] Kurt E. Sickafus, John M.Wills (1999) J Am Ceram Soc 82(12): 3279-3292. https://doi.org/10.1111/j.1151-2916.1999.tb02241.x

[30] Lakhani VK, Pathak TK, Vasoya NH, Modi KB (2011) Solid State Sci 13: 539-547. https://doi.org/10.1016/j.solidstatesciences.2010.12.023

[31] Li Q, Chang CB, Jing HX, Wang YF (2010) J Alloy Compd 495: 63-66. https://doi.org/10.1016/j.jallcom.2010.01.034

[32] Liu Q, Li L, Zhou JP, Chen XM, Bian XB, Liu P (2012) J Ceram Process Res 13:110-116.

[33] March A (1932) Z Kristallogr 81:285-297.

[34] MacGillavry CH, Rieck GD, Lonsdale K (1968) Physical and Chemical Tables, International Tables for X-Ray Crystallography Vol III Kynoch Press Birmingham UK.

[35] Momma K, Izumi F (2006) Comm Crystallogr Comput IUCr Newslett 7:106-119.

[36] Nasir S, Asghar G, Muhammad Ali Malik, Anis-ur-Rehman M (2011) J Sol Gel Sci Technol 59:111-116. https://doi.org/10.1007/s10971-011-2468-x

[37] Navneet Singh, Ashish Agarwal, Sujata Sanghi, Paramjeet Singh (2011) Physica B 406:687-692. https://doi.org/10.1016/j.physb.2010.11.087

[38] Neel (1948) Ann Phys Paris 3:137-198.

[39] Nikam DS, Jadhav SV, Khot VM, Bohara RA, Hong CK, Mali SS, Pawar SH (2015) RSC Adv 5:2338-2345. https://doi.org/10.1039/C4RA08342C

[40] Nikumbh A, Pawar RA, Nighot DV, Gugale GS, Sangale MD, Khanvilkar MB, Nagawade AV (2014) J Magn Magn Mater 355:201-209. https://doi.org/10.1016/j.jmmm.2013.11.052

[41] Nilar Lwin, Ahmad Fauzi Mohd Noor, Srimala Sreekantan, Radzali Othman (2013) Int J Appl Ceram Technol 10(6):1-7. https://doi.org/10.1111/ijac.12066

[42] O'Handley RC (2000) Modern magnetic materials: principles and applications John Wiley & Sons New York.

[43] Pancove J (1971) Optical process in semiconductors Prentice-Hall Englewood Cliffs NJ USA.

[44] Patange SM, Shirsath SE, Jadhav SS, Jadhav KM (2012) Phys Status Solidi A 209: 347-352. https://doi.org/10.1002/pssa.201127232

[45] Pereira FMM, Sombra ASB (2013) Solid State Phenom (2013) 202:1-64. https://doi.org/10.4028/www.scientific.net/SSP.202.1

[46] Perkampus, Heinz-Helmut, Translated by Charlotte Grinter H and Dr. Threlfall TL (1992) UV-VIS Spectroscopy and Its Applications Springer Laboratory Springer – Verlag.

[47] Petřiček V Dušek M, Palatinus L (2006) JANA2006: The Crystallographic Computing System, Institute of Physics, Academy of Sciences of the Czech Republic Praha 2000.

[48] Raghavan V (2012) Materials Science and Engineering 5th edition PHI Learning private limited New Delhi.

[49] Rangamohan G, Ravinder D, Ramanareddy AV, Boyanov BS (1999) Mater Lett 40:39-45. https://doi.org/10.1016/S0167-577X(99)00046-4

[50] Ravinder D, Rangamohan G, Prankishan, Nitendarkishan, Sagar DR (2000) Mater Lett 44: 256-260. https://doi.org/10.1016/S0167-577X(00)00039-2

[51] Reddy MP, Kim G, Yoo DS, Madhuri W, Reddy NR, Siva Kumar KV, Reddy RR (2012) Mater Sci Appl 3:628-632.

[52] Rietveld HM (1969) J Appl Crystallogr 2:65-71. https://doi.org/10.1107/S0021889869006558

[53] Ron Jenkins (1999) X-Ray Fluorescence Spectrometry Second Edition John Wiley & Sons. https://doi.org/10.1002/9781118521014

[54] Roshan L, Suman, Sharma ND, Taneja SP, Reddy VR (2007) Ind J Pure Appl Phy 45: 231-237.

[55] Sagar ES, Mahesh LM, Yukiko Y, Xiaoxi L, Akimitsu M (2014) Phys Chem Chem Phys 16:2347-2357. https://doi.org/10.1039/C3CP54257B

[56] Sakata M, Takata M (1996) High Press Res 14:327-333. https://doi.org/10.1080/08957959608201418

[57] Sandervan S, Jeanette N (2009) Phys Scr 79 048304 (9pp).

[58]   Saravanakumar S, Pattammal M, Israel S, Sheeba RAJR, Saravanan R (2012) Physica B 407:302-310. https://doi.org/10.1016/j.physb.2011.10.004

[59]   Saravanan R GRAIN software (private communication)

[60]   Saravanan R (2009) Phys Scr 79 048303 (8pp).

[61]   Saravanan R, Premarani M (2007) J Phys Condens Matter 19:266221. https://doi.org/10.1088/0953-8984/19/26/266221

[62]   Saravanan R, Premarani M (2012) Metal and Alloy Bonding: An Experimental Analysis, Charge Density in Metals and Alloys Springer-Verlag. https://doi.org/10.1007/978-1-4471-2204-3

[63]   Saravanan R, Santhanam F, Berchmans JL (2014) Chem Pap 68(6):788-797.

[64]   Shinde NS, Khot SS, More RM, Kale BB, Basavaiah N, Watawe SC, Vaidya MM (2014) Int J Chem Phys Sci ISSN:2319-6602 Vol 3 Special Issue – NCRTSM.

[65]   Sonali L.Darshane, Suryavanshi SS, Mulla IS (2009) Ceram Int 35:1793–1797. https://doi.org/10.1016/j.ceramint.2008.10.013

[66]   Standley KJ (1962) Oxide magnetic materials Oxford Clarendon Press.

[67]   Suryanarayana C (2001) Prog Mater Sci 46:1-184. https://doi.org/10.1016/S0079-6425(99)00010-9

[68]   Suryanarayana C, Grant NM (1998) X-Ray Diffraction A Practical Approach Springer Science Business Media LLC.

[69]   Takata M (2008) Acta Crystallogr Sect A 64:232-245. https://doi.org/10.1107/S010876730706521X

[70]   Tauc J, Grigorovic R, Vancu A (1966) Phys Status Solidi (b) 15:627-637. https://doi.org/10.1002/pssb.19660150224

[71]   Thomas JJ, Shinde AB, Krishna PSR, Kalarikka N (2013) J Alloys Compd 546:77-83. https://doi.org/10.1016/j.jallcom.2012.08.011

[72]   Tony O (1996) Fundamentals of Modern UV-Visible Spectroscopy: A Primer Hewlett-Packard Company Germany.

[73]   Viswanathan B, Murthy VRK (1990) Ferrite Materials Narosa publications New Delhi.

[74]   Wertheim GK, Butler MA, West KW, Buchanan DNE (1974) Rev Sci Instrum 45:1369-1371. https://doi.org/10.1063/1.1686503

[75]   Wood DL, Tauc J (1972) Phys Rev B 5:3144-3151.
       https://doi.org/10.1103/PhysRevB.5.3144

Chapter 2

Results

Abstract

Chapter 2 displays the results of the powder X-ray diffraction, scanning electron microscope, energy dispersive X-Ray analysis, X-Ray fluorescence, magnetic, optical and dielectric characterizations carried out on $NiFe_2O_4$, $Ni_{0.5}Zn_{0.5}Fe_2O_4$, $Ni_{0.8}Zn_{0.2}Fe_2O_4$, $Mn_{0.4}Zn_{0.6}Fe_2O_4$, $Co_{0.4}Zn_{0.6}Fe_2O_4$, $Ni_{0.53}Cu_{0.12}Zn_{0.35}Fe_2O_4$ and $Mg_{0.2}Cu_{0.3}Zn_{0.5}Fe_2O_4$ ferrite samples synthesized for this research work. The results of fitted PXRD profiles for all the prepared nano ferrite materials are reported. Particle size evaluation from SEM, crystallite sizes from powder X-ray profile, elemental compositional analyses of the synthesized nano ferrite materials are also presented. Electron density distribution studies are carried out using the maximum entropy method. The results of the electron density distribution studies are presented in the form of three, two and one dimensional electron density maps.

Keywords

Powder XRD, X-ray Fluorescence, SEM, EDX, Electron Density, TGA, DTA, Dielectric

Contents

2.   Introduction...43

2.1   Structural characterization of the synthesized samples 43

2.2   Results – Raw powder XRD data..44

2.3   Results – Cation distribution 46

2.4   Results – Refined powder XRD of synthesized samples 48

2.5   TG/DTA analysis of $Ni_{0.5}Zn_{0.5}Fe_2O_4$ (700 - 1000°C)
        and $Ni_{0.8}Zn_{0.2}Fe_2O_4$ samples 64

2.6   Morphological studies of synthesized samples...................................67

2.6.1   SEM results 67

2.6.2   EDAX and XRF characterization results..75

2.7   Results – MEM analysis 81

2.8   Results - Magnetic properties **105**

2.9   Results - Optical properties **111**

2.10   Results - Dielectric properties **115**

References **119**

2.   Introduction

The synthesized nano ferrite samples were analyzed by various characterization techniques with the intention of validating the accomplishment made on desired phases/qualities of the samples. Structural characterization of the synthesized samples was done with the aid of X-ray diffraction data. The morphology of the samples was analyzed by scanning electron microscope (SEM) and the elemental composition of the samples was studied by energy dispersive X-ray spectroscopic analysis (EDX) and X-ray fluorescence (XRF) studies. The magnetic properties, optical properties and dielectric properties were also investigated by vibrating sample magnetometer (VSM), UV-Vis spectrometer and broad band dielectric spectrometer studies respectively. The results are presented in this chapter.

2.1   Structural characterization of the synthesized samples

The synthesized samples were subjected to the non-destructive X-ray diffraction study and from resultant X-ray diffraction results, the formation of required phases in the samples was ascertained by comparing the experimental Bragg peaks with those of the Joint Committee on Powder Diffraction Standards (JCPDS) cards of the respective sample. The intensity and width of various Bragg peaks in the raw X-ray diffraction profile indicate the size of the particles and distribution of cations in the unit cell of ferrites (Gangatharan et al., 2010; Deraz et al., 2012; Dalibor et al., 2015). The raw X-ray diffraction data were refined by the Rietveld method (Rietveld, 1969) using the software JANA 2006 (Petříček et al., 2006). From the refinement results, the structural parameters like, lattice parameter (a_{exp}), oxygen positional parameter (u) etc., can be obtained. The values of 'a_{exp}' and 'u' can be utilized in evaluating the radii of A-site and B-site ($r_{A(XRD)}$, $r_{B(XRD)}$), various inter atomic distances between cation – anion (M_e – O), cation – cation (M_e – M_e) and inter atomic bond angles (θ_1 – θ_5). The 'a_{exp}' and 'u' values are the outcome of the distribution of cations in the lattice and hence, any deviation in a_{exp} and u values, due to the difference in radii values of substituted divalent ions, reflects in the spin interaction between the A-site and B-site ions (Satalkar et al., 2016) which in turn determines the mid bond electron density value between the two sites. The grain size is evaluated using the software GRAIN (Saravanan R, GRAIN software), because it is one of the major factors influencing the magnetic properties of ferrites (Sidra et al., 2016).

2.2   Results – Raw powder XRD data

The raw X-ray diffractogram of the synthesized samples are shown in Figs. 2.1(a–e).

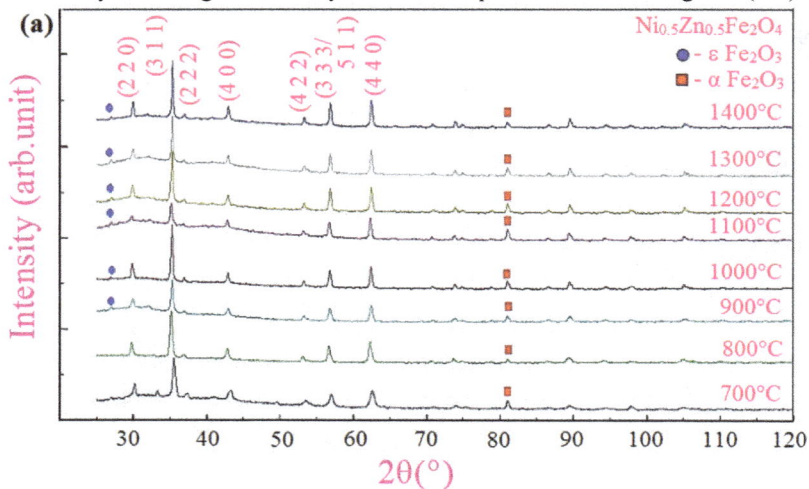

Fig.2.1(a) *Raw XRD powder profiles of $Ni_{0.5}Zn_{0.5}Fe_2O_4$ samples sintered from 700°C to 1400°C.*

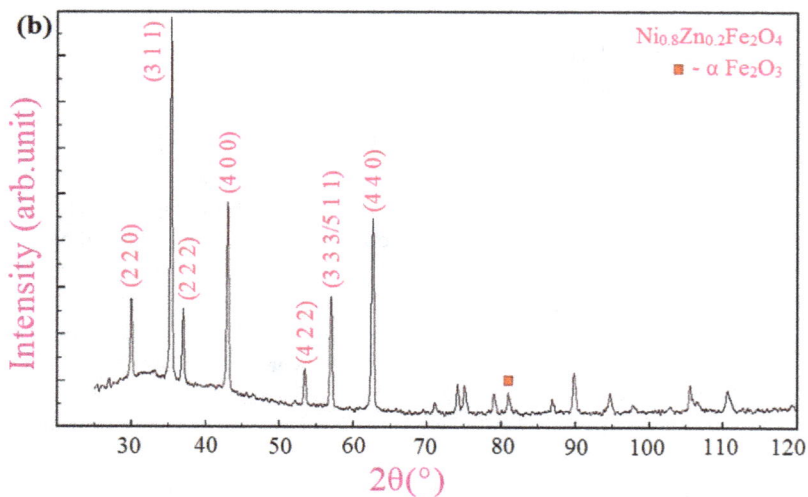

Fig.2.1(b) *Raw XRD powder profile of $Ni_{0.8}Zn_{0.2}Fe_2O_4$ sample sintered at 900°C.*

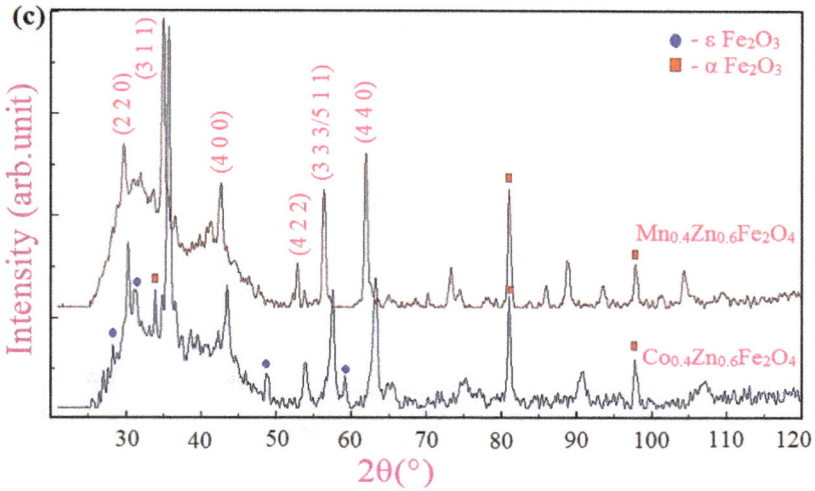

Fig.2.1(c) *Raw XRD powder profiles of $Mn_{0.4}Zn_{0.6}Fe_2O_4$ and $Co_{0.4}Zn_{0.6}Fe_2O_4$ samples sintered at $900\,°C$.*

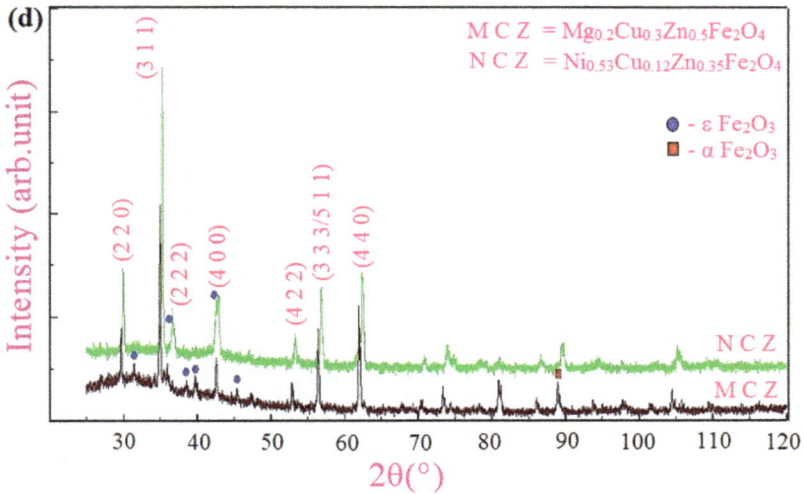

Fig.2.1(d) *Raw XRD powder profiles of $Ni_{0.53}Cu_{0.12}Zn_{0.35}Fe_2O_4$ and $Mg_{0.2}Cu_{0.3}Zn_{0.5}Fe_2O_4$ samples sintered at $900\,°C$.*

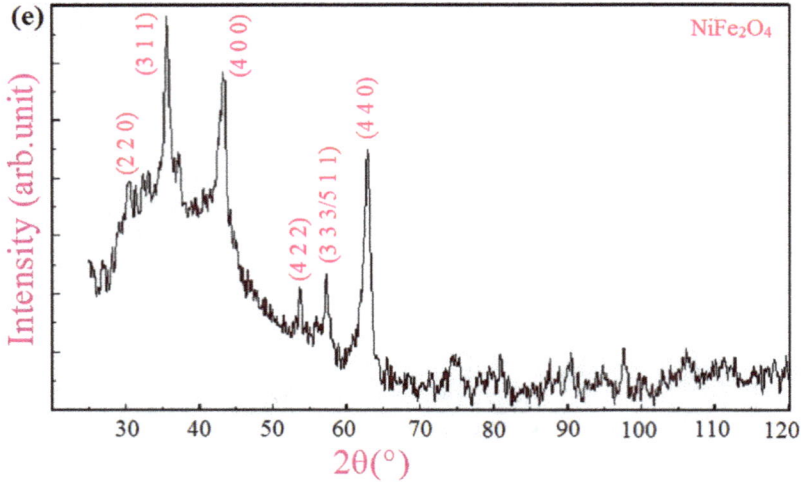

Fig.2.1(e) *Raw XRD powder profile of NiFe$_2$O$_4$ sample prepared by chemical reaction technique and calcination done at 700°C.*

2.3 Results – Cation distribution

The distribution of cation in the unit cell among the A-site and B-site is determined using Eqs.(1.24 - 1.31), as explained in sec. 1.11.1. The X-ray intensity ratios of I_{220}/I_{400}, I_{220}/I_{440}, I_{422}/I_{440} and I_{400}/I_{440} are found out for various cation distribution combinations and those combinations that match closely with observed intensity ratios are considered as cation distributions. The intensity ratio of various Bragg planes and the cation distribution, obtained in the present research study are tabulated in Tables 2.1 and 2.2 respectively. The correlation between the intensities of (220) plane (I_{220}) and the content of Fe ions at A-site for Ni$_{0.5}$Zn$_{0.5}$Fe$_2$O$_4$ samples sintered at various temperatures from 700 -1400°C is shown in Fig.2.2 and the same for the samples Mn$_{0.4}$Zn$_{0.6}$Fe$_2$O$_4$ and Co$_{0.4}$Zn$_{0.6}$Fe$_2$O$_4$ is tabulated in Table 2.3.

Table 2.1 *X-ray intensity ratio of synthesized samples.*

Sample	Temp (°C)	I_{220}/I_{400}		I_{220}/I_{440}		I_{422}/I_{440}		I_{400}/I_{440}	
		Obs	Cal	Obs	Cal	Obs	Cal	Obs	Cal
	700	1.54	1.34	0.87	1.08	0.33	0.33	0.54	0.80
	800	1.33	1.41	0.70	1.13	0.29	0.35	0.53	0.80
	900	1.33	1.42	0.67	1.13	0.31	0.35	0.50	0.79
$Ni_{0.5}Zn_{0.5}Fe_2O_4$	1000	1.35	1.47	0.71	1.16	0.30	0.35	0.53	0.78
	1100	0.85	1.36	0.36	1.09	0.26	0.34	0.42	0.81
	1200	1.47	1.47	0.63	1.15	0.28	0.35	0.43	0.79
	1300	1.34	1.51	0.45	1.17	0.24	0.36	0.34	0.78
	1400	1.15	1.49	0.63	1.16	0.29	0.36	0.55	0.78
$Ni_{0.8}Zn_{0.2}Fe_2O_4$	900	0.44	0.30	0.43	0.24	0.19	0.09	0.98	0.82
$Mn_{0.4}Zn_{0.6}Fe_2O_4$	900	0.91	1.35	0.40	1.10	0.28	0.33	0.44	0.82
$Co_{0.4}Zn_{0.6}Fe_2O_4$	900	1.39	1.39	0.84	1.20	0.34	0.36	0.61	0.87
$Ni_{0.53}Cu_{0.12}Zn_{0.35}Fe_2O_4$	900	1.29	0.40	0.84	0.84	0.28	0.32	0.65	2.08
$Mg_{0.2}Cu_{0.3}Zn_{0.5}Fe_2O_4$	900	1.20	0.34	0.52	0.52	0.26	0.20	0.43	1.56
$NiFe_2O_4$	700	0.27	0.27	0.20	0.21	0.29	0.13	0.76	0.80

Table 2.2 *Cation distribution of synthesized samples.*

Sample	Temp (°C)	Cation distribution A – site	B – site
	700	Fe_1	$Zn_{0.5}Ni_{0.5}Fe_1$
	800	$Zn_{0.075}Ni_{0.054}Fe_{0.871}$	$Zn_{0.425}Ni_{0.446}Fe_{1.129}$
	900	$Zn_{0.078}Ni_{0.082}Fe_{0.84}$	$Zn_{0.422}Ni_{0.418}Fe_{1.16}$
$Ni_{0.5}Zn_{0.5}Fe_2O_4$	1000	$Zn_{0.086}Ni_{0.210}Fe_{0.704}$	$Zn_{0.414}Ni_{0.290}Fe_{1.296}$
	1100	$Ni_{0.045}Fe_{0.955}$	$Zn_{0.5}Ni_{0.455}Fe_{1.045}$
	1200	$Zn_{0.1827}Fe_{0.8173}$	$Zn_{0.3173}Ni_{0.5}Fe_{1.1827}$
	1300	$Zn_{0.233}Fe_{0.767}$	$Zn_{0.267}Ni_{0.5}Fe_{1.233}$
	1400	$Zn_{0.218}Fe_{0.782}$	$Zn_{0.282}Ni_{0.5}Fe_{1.218}$
$Ni_{0.8}Zn_{0.2}Fe_2O_4$	900	Fe_1	$Ni_{0.8}Zn_{0.2}Fe_1$
$Mn_{0.4}Zn_{0.6}Fe_2O_4$	900	$Mn_{0.4}Fe_{0.6}$	$Zn_{0.6}Fe_{1.4}$
$Co_{0.4}Zn_{0.6}Fe_2O_4$	900	$Zn_{0.245}Co_{0.04}Fe_{0.715}$	$Zn_{0.355}Co_{0.36}Fe_{1.285}$
$Ni_{0.53}Cu_{0.12}Zn_{0.35}Fe_2O_4$	900	$Ni_{0.2888}Cu_{0.04}$ $Zn_{0.3043}Fe_{0.3669}$	$Ni_{0.2412}Zn_{0.0457}$ $Cu_{0.08}Fe_{1.6331}$
$Mg_{0.2}Cu_{0.3}Zn_{0.5}Fe_2O_4$	900	$Mg_{0.0405}Cu_{0.15}$ $Zn_{0.5}Fe_{0.3095}$	$Mg_{0.1595}Cu_{0.15}$ $Fe_{1.6905}$
$NiFe_2O_4$	700	$Ni_{0.065}Fe_{0.935}$	$Ni_{0.935}Fe_{1.065}$

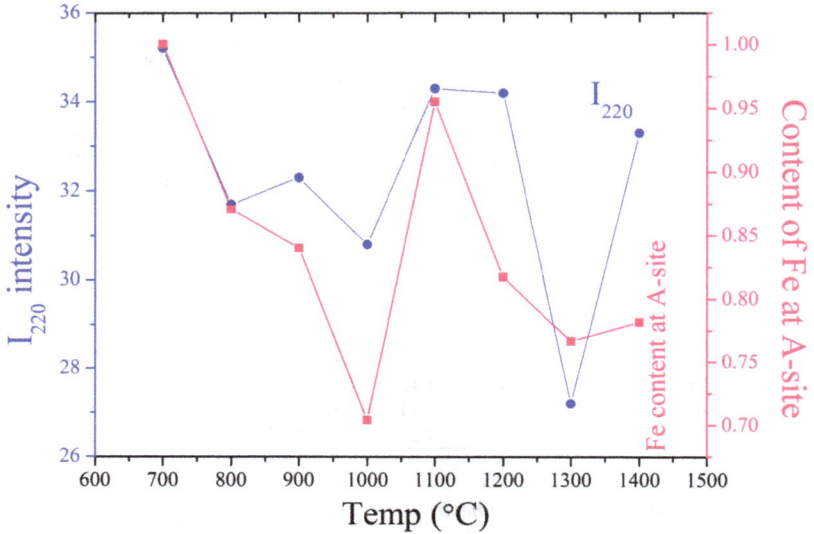

Fig.2.2 *Correlation between I_{220} intensity values and content of Fe ions at A-site for $Ni_{0.5}Zn_{0.5}Fe_2O_4$ samples at various tempertures.*

Table 2.3 *Correlation between I_{220} intensity and content of Fe at A-site of $Mn_{0.4}Zn_{0.6}Fe_2O_4$ and $Co_{0.4}Zn_{0.6}Fe_2O_4$ nano ferrite samples.*

Sample	I_{220}	Content of Fe at A-site
$Mn_{0.4}Zn_{0.6}Fe_2O_4$	28.5	0.600
$Co_{0.4}Zn_{0.6}Fe_2O_4$	29.6	0.715

2.4   Results – Refined powder XRD of synthesized samples

The cation distribution of the respective samples, as tabulated in Table 2.2, is employed in the Rietveld refinement method (Rietveld, 1969) using the software JANA 2006 (Petříček *et al.,* 2006), to refine the raw X-ray diffractograms of the synthesized samples. The resultant refined X-ray diffractograms of $Ni_{0.5}Zn_{0.5}Fe_2O_4$ samples sintered at various temperatures from 700 to 1400°C, $Ni_{0.8}Zn_{0.2}Fe_2O_4$, $Mn_{0.4}Zn_{0.6}Fe_2O_4$, $Co_{0.4}Zn_{0.6}Fe_2O_4$, $Ni_{0.53}Cu_{0.12}Zn_{0.35}Fe_2O_4$, $Mg_{0.2}Cu_{0.3}Zn_{0.5}Fe_2O_4$ and $NiFe_2O_4$ are shown in Figs.2.3(a–n) and the obtained reliability indices are tabulated in Table 2.4. The analysis that distinguishes this work from others is the evaluation of mid bond electron density between A-A, A-B and B-B site bonds in the unit cell of ferrite, determined by maximum entropy method (Gull and Daniel, 1978). The structure factor values, both observed (F_o) and calculated (F_c), considered for the evaluation of mid bond electron density are obtained from the Rietveld refinement method (Rietveld, 1969) using the software JANA 2006 (Petříček et al., 2006) and are given in Tables 2.5(a-c). The experimental lattice parameter (a_{exp}) and oxygen positional parameter (u) are also obtained from the results of the Rietveld refinement method (Rietveld, 1969) again using the software JANA 2006 (Petříček et al., 2006) and are tabulated in Table 2.6 along with A-site and B-site radii ($r_{A(XRD)}$, $r_{B(XRD)}$) evaluated from Eq.(1.32) and Eq.(1.33) respectively. Shared edge length of A and B-site (d_{AES}, d_{BES}) are calculated using Eq.(1.34) and Eq.(1.35) respectively and unshared edge length of B-site (d_{BEU}) calculated using Eq.(1.36). A comparison between 'a_{exp}' of the synthesized samples and lattice parameter reported in JCPDS of the respective samples is listed in Table 2.7. Fig.2.4 depicts the relation between the 'a_{exp}' and $r_{A(XRD)}$. The volume of the unit cell ($V = a_{exp}^3$) is evaluated using a_{exp} and X-ray density (D_x) and average grain size (t) of the samples synthesized in the present research work are evaluated using Eq.(1.37) and Eq.(1.38) respectively and tabulated in Table 2.8. The radii of A-site and B-site ($r_{A(CD)}$, $r_{B(CD)}$) are estimated from the knowledge of cation distribution using Eq.(1.39) and Eq.(1.40) respectively and lattice parameter (a_{the}) have been estimated theoretically from $r_{A(CD)}$ and $r_{B(CD)}$ values using Eq.1.41 and tabulated in Table 2.9.

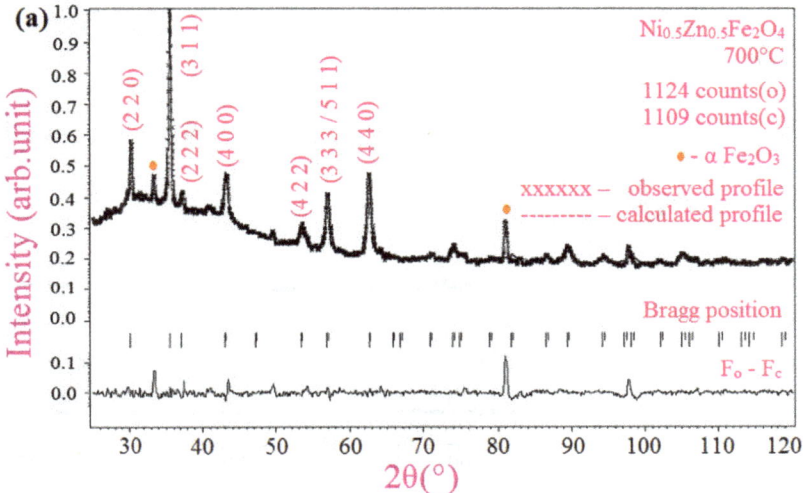

Fig.2.3(a) Fitted powder XRD profile of $Ni_{0.5}Zn_{0.5}Fe_2O_4$ nano ferrites sintered at 700°C.

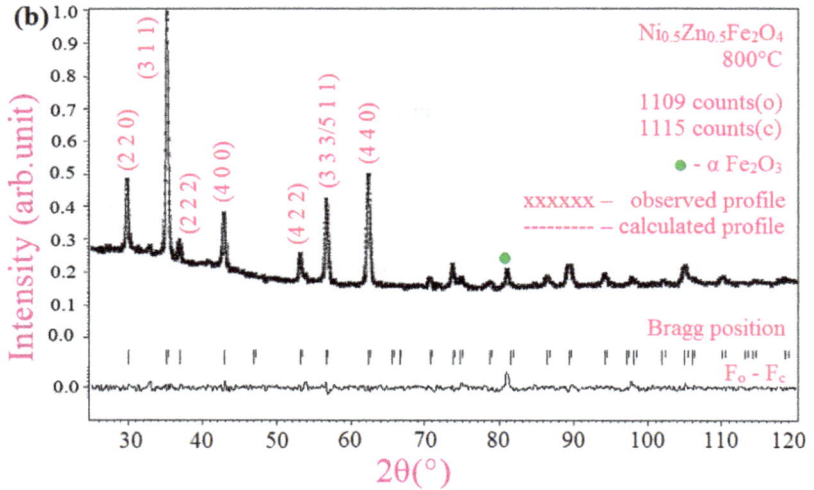

Fig.2.3(b) *Fitted powder XRD profile of $Ni_{0.5}Zn_{0.5}Fe_2O_4$ nano ferrites sintered at 800°C.*

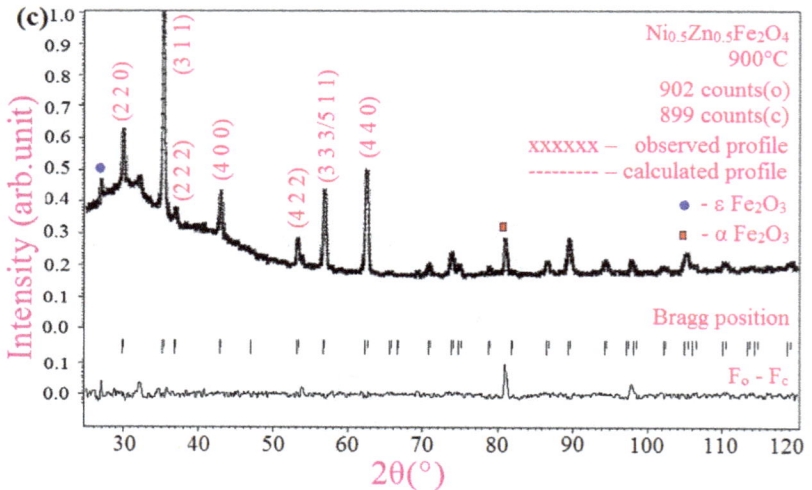

Fig.2.3(c) *Fitted powder XRD profile of $Ni_{0.5}Zn_{0.5}Fe_2O_4$ nano ferrites sintered at 900°C.*

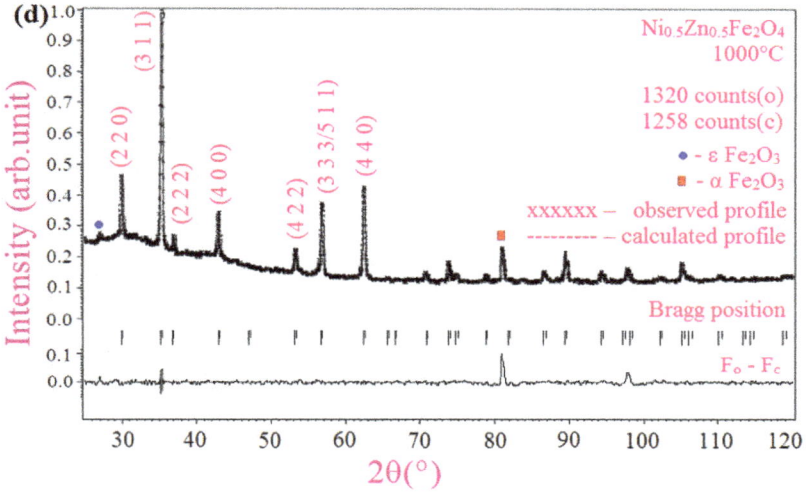

Fig.2.3(d) *Fitted powder XRD profile of* $Ni_{0.5}Zn_{0.5}Fe_2O_4$ *nano ferrites sintered at* 1000°C.

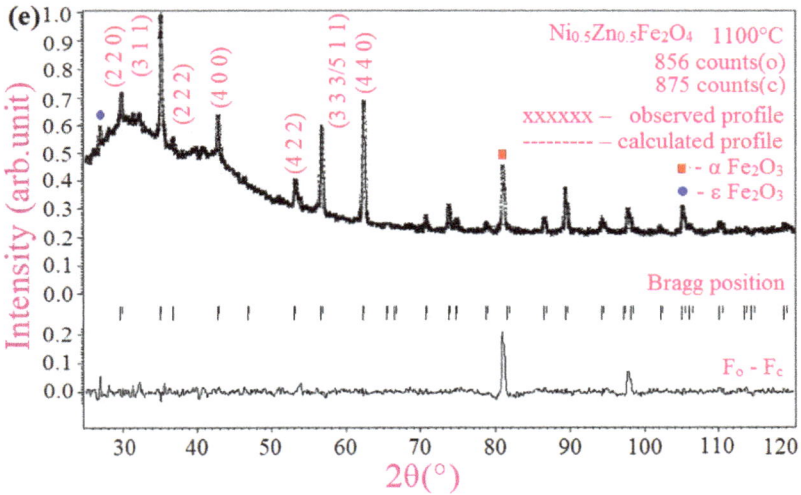

Fig.2.3(e) *Fitted powder XRD profile of* $Ni_{0.5}Zn_{0.5}Fe_2O_4$ *nano ferrites sintered at* 1100°C.

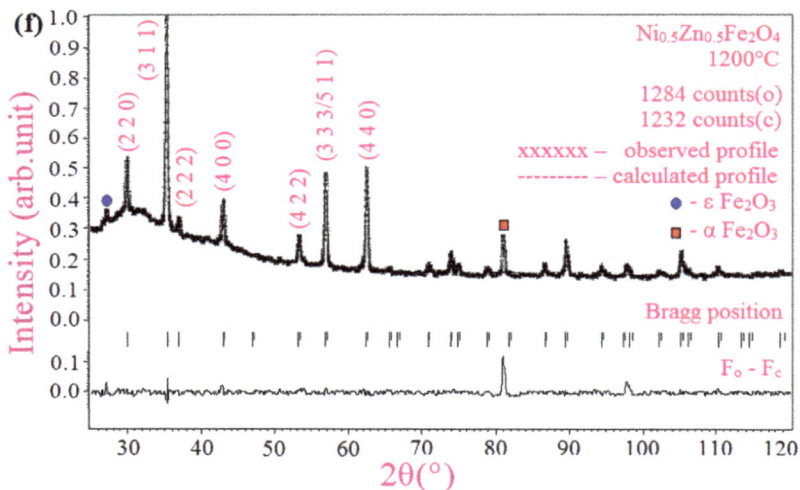

Fig.2.3(f) *Fitted powder XRD profile of $Ni_{0.5}Zn_{0.5}Fe_2O_4$ nano ferrites sintered at 1200°C.*

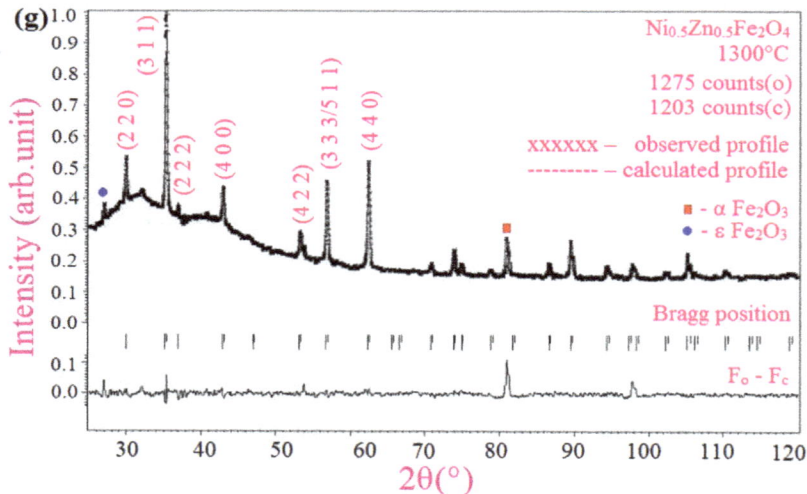

Fig.2.3(g) *Fitted powder XRD profile of $Ni_{0.5}Zn_{0.5}Fe_2O_4$ nano ferrites sintered at 1300°C.*

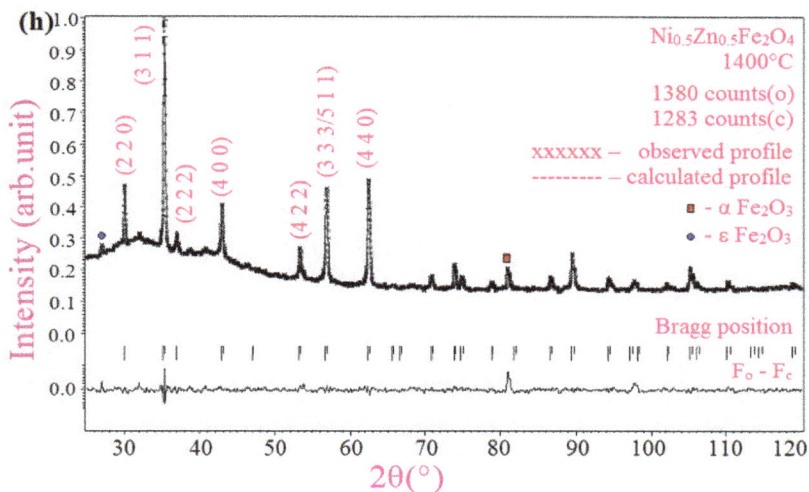

Fig.2.3(h) *Fitted powder XRD profile of $Ni_{0.5}Zn_{0.5}Fe_2O_4$ nano ferrites sintered at 1400°C.*

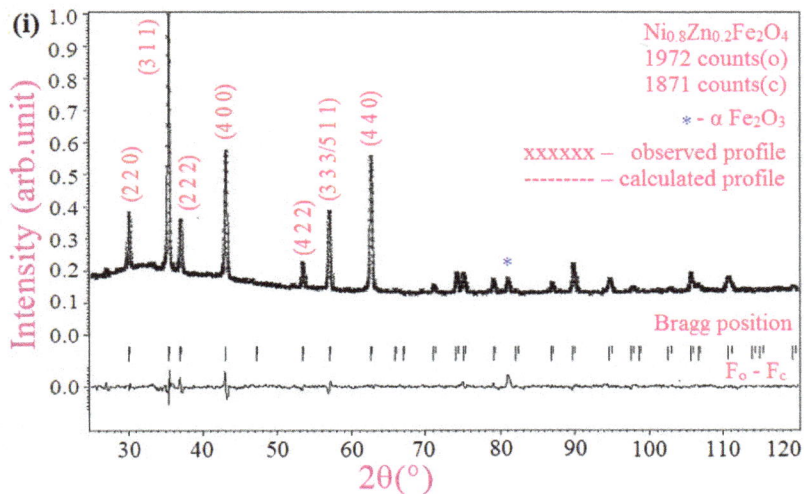

Fig.2.3(i) *Fitted powder XRD profile of $Ni_{0.8}Zn_{0.2}Fe_2O_4$ nano ferrites sintered at 900°C.*

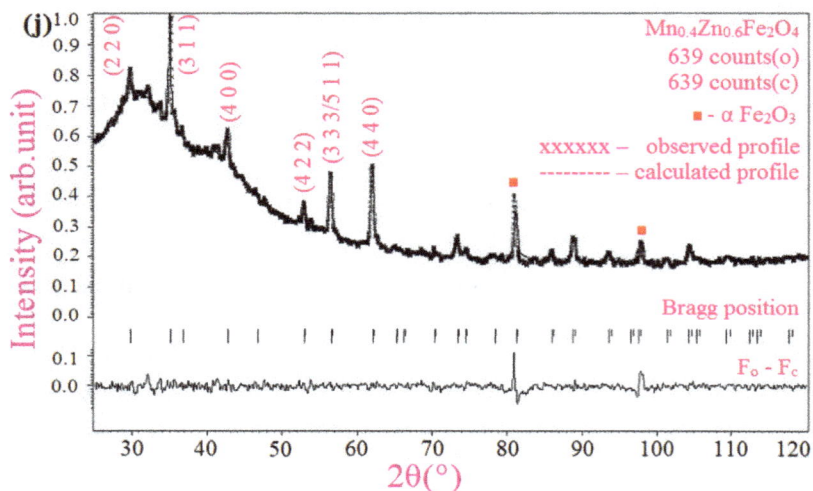

Fig.2.3(j) *Fitted powder XRD profile of $Mn_{0.4}Zn_{0.6}Fe_2O_4$ nano ferrites sintered at 900°C.*

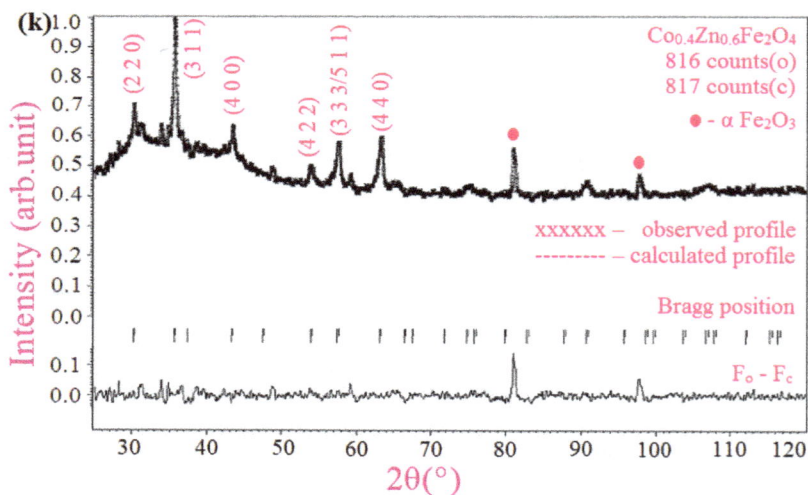

Fig.2.3(k) *Fitted powder XRD profile of $Co_{0.4}Zn_{0.6}Fe_2O_4$ nano ferrites sintered at 900°C.*

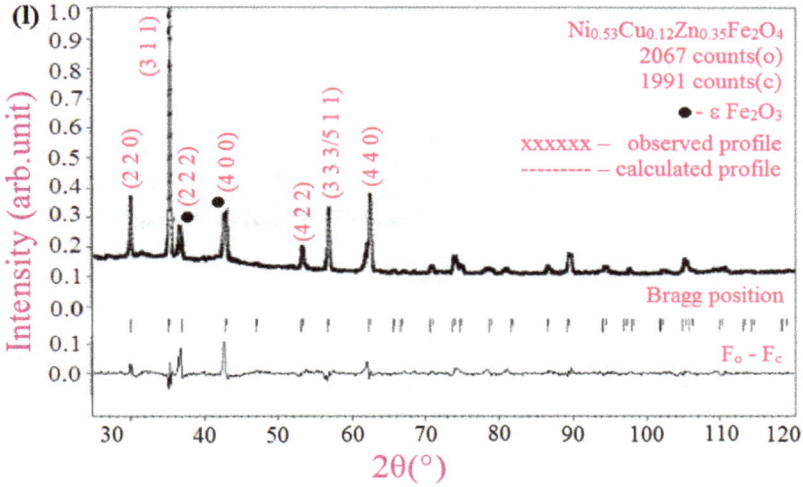

Fig.2.3(l) *Fitted powder XRD profile of $Ni_{0.53}Cu_{0.12}Zn_{0.35}Fe_2O_4$ nano ferrites sintered at 900°C.*

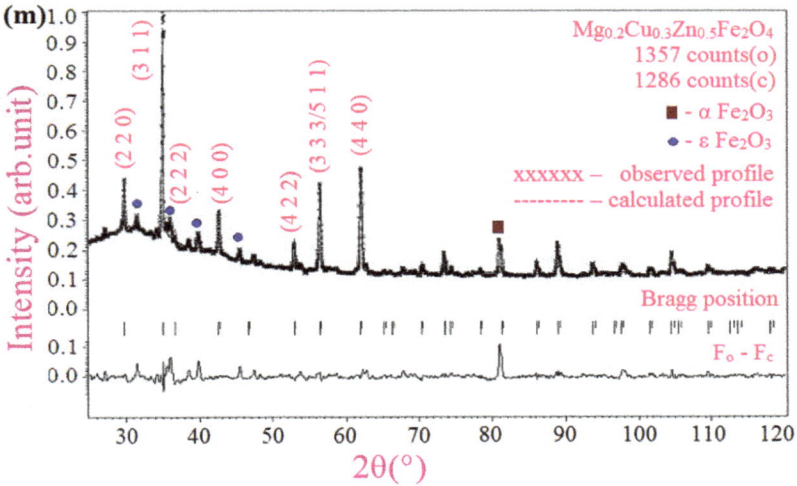

Fig.2.3(m) *Fitted powder XRD profile of $Mg_{0.2}Cu_{0.3}Zn_{0.5}Fe_2O_4$ nano ferrites sintered at 900°C.*

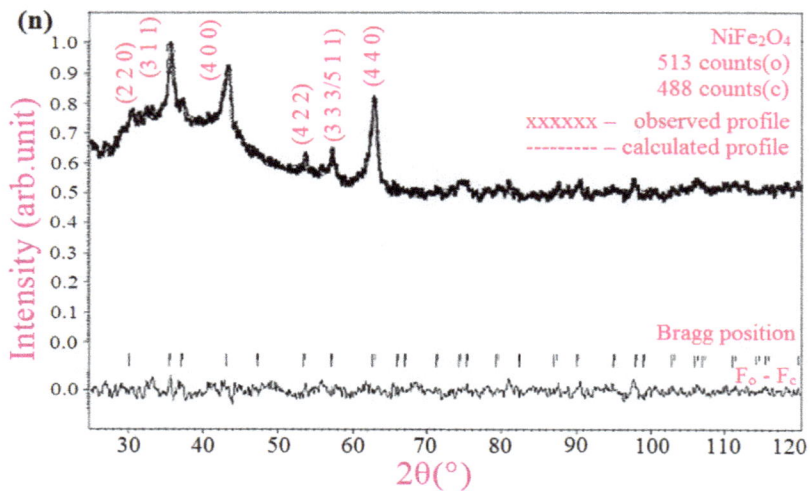

Fig.2.3(n) *Fitted powder XRD profile of NiFe$_2$O$_4$ nano ferrites calcined at 700°C.*

Table 2.4 Reliability indices obtained from Rietveld refinement method (Rietveld, 1969) using JANA 2006 (Petříček et al., 2006) software for samples synthesized in the present research work.

Sample	Temp(°C)	R_{obs}(%)	wR_{obs}(%)	R_{all}(%)	wR_{all}(%)	R_p(%)	wR_p(%)	GOF
	700	2.94	2.30	4.13	2.68	2.32	3.88	0.65
	800	2.24	2.12	2.63	2.29	1.99	2.77	0.41
	900	3.17	2.55	5.47	4.51	2.08	3.35	0.49
	1000	1.91	1.50	2.52	1.70	2.13	3.95	0.58
$Ni_{0.5}Zn_{0.5}Fe_2O_4$	1100	2.57	2.03	4.76	2.97	2.21	4.25	0.71
	1200	2.31	1.88	4.01	2.60	2.16	3.84	0.62
	1300	2.98	2.81	5.71	3.14	2.06	3.63	0.61
	1400	4.13	4.70	4.78	4.86	2.08	3.06	0.49
$Ni_{0.8}Zn_{0.2}Fe_2O_4$	900	9.45	11.27	10.72	11.54	2.28	3.27	0.59
$Mn_{0.4}Zn_{0.6}Fe_2O_4$	900	6.58	4.72	11.11	5.98	2.47	4.41	0.63
$Co_{0.4}Zn_{0.6}Fe_2O_4$	900	2.44	2.23	11.47	5.59	1.82	2.74	0.53
$Ni_{0.53}Cu_{0.12}Zn_{0.35}Fe_2O_4$	900	6.18	4.56	6.96	4.75	3.21	5.35	0.91
$Mg_{0.2}Cu_{0.3}Zn_{0.5}Fe_2O_4$	900	3.87	3.51	8.21	4.21	3.76	5.83	0.85
$NiFe_2O_4$	700	5.90	5.44	9.07	6.65	1.97	2.49	0.43

R_{obs}(%) - Reliability index, where $R_{obs} = \frac{\sum||F_o|-|F_c||}{\sum|F_o|}$

wR_{obs}(%) - Weighted reliability index, where $wR_{obs} = \sqrt{\frac{\sum[w(|F_o|^2-|F_c|^2)^2]}{\sum[w(|F_o|^4)]}}$

R_p(%) - Profile reliability index, where $R_p = \frac{\sum|I_o-I_c|}{\sum|I_o|}$

wR_p(%) - Weighted profile reliability index, where $wR_p = \sqrt{\frac{\sum[w(|I_o|-|I_c|)^2]}{\sum[w(|I_o|^2)]}}$

GOF - Goodness of fit $= \frac{wR_p}{R_{exp}}$ where $R_{exp} = \left[\frac{N-P}{\sum[w(|I_o|^2)]}\right]^{1/2}$ is the expected error.

N – P = number of observations – number of least-square parameters.

Table 2.5(a) *Observed (F_o) and calculated (F_c) structure factor obtained from the Rietveld refinement method (Rietveld, 1969) using the JANA 2006 software (Petříček et al., 2006) for $Ni_{0.5}Zn_{0.5}Fe_2O_4$ samples sintered at different temperatures.*

h k l	$Ni_{0.5}Zn_{0.5}Fe_2O_4$							
	700°C		800°C		900°C		1000°C	
	F_o	F_c	F_o	F_c	F_o	F_c	F_o	F_c
2 2 0	160.33	161.35	161.02	160.64	161.81	165.63	158.88	162.29
3 1 1	257.98	257.45	256.27	255.68	262.78	264.17	252.92	252.81
2 2 2	132.52	142.68	129.37	130.26	129.26	136.47	131.77	134.22
4 0 0	252.07	255.37	291.80	289.26	289.56	285.90	272.68	271.48
4 2 2	126.75	127.49	120.55	120.80	132.21	133.58	120.81	120.77
5 1 1	191.87	193.65	206.02	208.17	214.63	218.21	208.39	208.00
4 4 0	387.07	388.06	419.64	422.11	444.04	447.00	399.02	400.22
h k l	1100°C		1200°C		1300°C		1400°C	
	F_o	F_c	F_o	F_c	F_o	F_c	F_o	F_c
2 2 0	157.33	162.48	176.08	175.21	182.68	177.91	180.57	178.76
3 1 1	249.39	254.52	262.09	263.38	277.44	276.76	264.22	265.97
2 2 2	132.37	135.48	138.36	144.19	117.45	141.27	138.21	150.63
4 0 0	298.39	284.53	287.70	286.37	292.51	302.07	284.08	282.33
4 2 2	122.56	122.86	145.24	136.01	151.97	152.47	147.38	140.32
5 1 1	217..89	221.26	259.16	258.04	263.78	264.52	266.88	266.38
4 4 0	418.35	420.79	462.93	465.60	529.09	523.94	468.28	472.63

Table 2.5(b) Observed (F_o) and calculated (F_c) structure factor obtained from the Rietveld refinement method (Rietveld, 1969) using the JANA 2006 software (Petříček *et al.,* 2006) for $Ni_{0.8}Zn_{0.2}Fe_2O_4$, $Mn_{0.4}Zn_{0.6}Fe_2O_4$ and $Co_{0.4}Zn_{0.6}Fe_2O_4$ samples

h k l	$Ni_{0.8}Zn_{0.2}Fe_2O_4$		$Mn_{0.4}Zn_{0.6}Fe_2O_4$		$Co_{0.4}Zn_{0.6}Fe_2O_4$	
	F_o	F_c	F_o	F_c	F_o	F_c
2 2 0	150.17	152.30	163.62	166.88	171.30	172.94
3 1 1	253.12	253.78	246.72	245.61	262.45	261.82
2 2 2	203.99	191.01	142.57	151.49	67.40	116.18
4 0 0	260.03	256.75	311.17	299.57	281.70	302.38
4 2 2	114.08	106.29	122.16	103.61	153.18	136.45
5 1 1	196.55	201.52	242.72	250.06	248.38	244.53
4 4 0	361.78	367.07	409.97	414.44	482.22	486.71

Table 2.5(c) Observed (F_o) and calculated (F_c) structure factor obtained from the Rietveld refinement method (Rietveld, 1969) using the JANA 2006 software (Petříček *et al.,* 2006) for $Ni_{0.53}Cu_{0.12}Zn_{0.35}Fe_2O_4$ and $Mg_{0.2}Cu_{0.3}Zn_{0.5}Fe_2O_4$ and $NiFe_2O_4$ samples.

h k l	$Ni_{0.53}Cu_{0.12}Zn_{0.35}Fe_2O_4$		$Mg_{0.2}Cu_{0.3}Zn_{0.5}Fe_2O_4$		$NiFe_2O_4$	
	F_o	F_c	F_o	F_c	F_o	F_c
2 2 0	179.61	170.78	186.5	182.4	105.79	106.51
3 1 1	256.99	263.84	259.7	260.7	196.45	203.77
4 0 0	295.55	295.86	268.1	267.0	364.48	367.71
4 2 2	130.85	130.74	138.2	148.9	41.75	35.82
5 1 1	224.20	234.13	243.8	245.7	132.02	141.82
4 4 0	464.98	466.05	479.9	474.9	375.16	345.63

Table 2.6 Structural parameters obtained from th eRietveld refinement method (Rietveld, 1969)

Sample	Temp (°C)	a_{exp} (Å)	u (Å)	$r_{A(XRD)}$ (Å)	$r_{B(XRD)}$ (Å)	d_{AES} (Å)	d_{BES} (Å)	d_{BEU} (Å)
	700	8.409(3)	0.380(2)	0.569	0.743	3.085	2.861	2.974
	800	8.398(2)	0.380(4)	0.566	0.740	3.080	2.858	2.970
	900	8.391(2)	0.378(1)	0.539	0.753	3.036	2.897	2.967
$Ni_{0.5}Zn_{0.5}Fe_2O_4$	1000	8.391(1)	0.382(2)	0.595	0.721	3.128	2.806	2.969
	1100	8.383(9)	0.383(5)	0.626	0.700	3.178	2.750	2.968
	1200	8.391(2)	0.389(4)	0.702	0.659	3.302	2.632	2.976
	1300	8.386(2)	0.382(8)	0.610	0.711	3.151	2.778	2.968
	1400	8.393(2)	0.390(4)	0.720	0.649	3.332	2.603	2.979
$Ni_{0.8}Zn_{0.2}Fe_2O_4$	900	8.364(8)	0.379(1)	0.555	0.734	3.062	2.852	2.958
$Mn_{0.4}Zn_{0.6}Fe_2O_4$	900	8.439(2)	0.396(3)	0.815	0.612	3.486	2.481	3.005
$Co_{0.4}Zn_{0.6}Fe_2O_4$	900	8.307(1)	0.384(1)	0.615	0.678	3.159	2.715	2.941
$Ni_{0.53}Cu_{0.12}Zn_{0.35}Fe_2O_4$	900	8.405(7)	0.382(8)	0.605	0.721	3.143	2.800	2.974
$Mg_{0.2}Cu_{0.3}Zn_{0.5}Fe_2O_4$	900	8.419(3)	0.384(7)	0.643	0.704	3.206	2.747	2.981
$NiFe_2O_4$	700	8.351(2)	0.379(4)	0.548	0.733	3.050	2.855	2.953

a_{exp} = experimental lattice parameter

u = Oxygen positional parameter

r_A and r_B = tetrahedral and octahedral site radii obtained from XRD data

d_{AES} = Tetrahedral shared edge length

d_{BES} = Octahedral shared edge length

d_{BEU} = Octahedral unshared edge length

Table 2.7 *Correlation of Bragg diffraction angle $(2\theta_{311})$ and lattice parameter value between reported in JCPDS and samples synthesized in the present research work.*

Sample	Temp (°C)	JCPDS $2\theta_{311}(°)$	a (Å)	Synthesized nano samples $2\theta_{311}(°)$	a_{exp}(Å)
$Ni_{0.5}Zn_{0.5}Fe_2O_4$	700			35.551	8.409(3)
	800			35.179	8.398(2)
	900			35.316	8.391(2)
	1000	35.569	8.382	35.272	8.391(1)
	1100	(JCDPS– 520278)		35.182	8.383(9)
	1200			35.318	8.391(2)
	1300			35.329	8.386(2)
	1400			35.357	8.393(2)
$Ni_{0.8}Zn_{0.2}Fe_2O_4$	900	35.715 (JCPDS – 441485)	8.339	35.432	8.364(8)
$Mn_{0.4}Zn_{0.6}Fe_2O_4$	900	35.294 (JCPDS – 221012)	8.441	35.048	8.439(2)
$Co_{0.4}Zn_{0.6}Fe_2O_4$	900	35.294 (JCPDS – 221012)	8.441	33.961	8.307(1)
$Ni_{0.53}Cu_{0.12}Zn_{0.35}Fe_2O_4$	900	35.730 (JCPDS – 100325)	8.339	35.333	8.405(7)
$Mg_{0.2}Cu_{0.3}Zn_{0.5}Fe_2O_4$	900	35.294 (JCPDS – 221012)	8.441	35.018	8.419(3)
$NiFe_2O_4$	700	35.730 (JCPDS – 100325)	8.339	35.625	8.351(2)

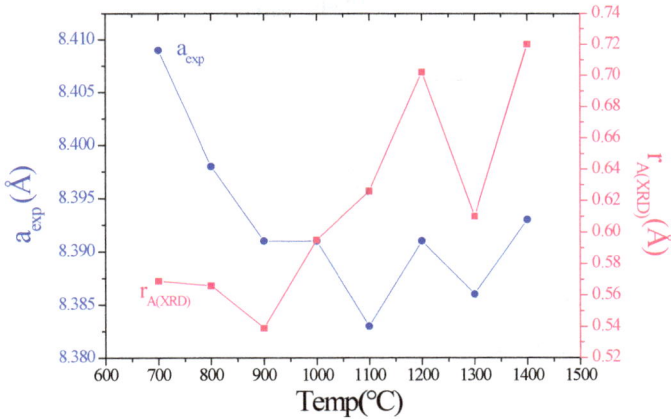

Fig.2.4 *Relation between experimental lattice parameter (a_{exp}) and radius of the A-site ($r_{A(XRD)}$) of $Ni_{0.5}Zn_{0.5}Fe_2O_4$ samples.*

Table 2.8 Cell volume, X-ray density and average crystallite size of synthesized samples.

Sample	Temp (°C)	V (Å)^3	D_x (g/cm^3)	t (nm)
Ni$_{0.5}$Zn$_{0.5}$Fe$_2$O$_4$	700	594.6	5.310	17(2)
	800	592.2	5.332	23(2)
	900	590.8	5.344	26(3)
	1000	590.8	5.345	28(1)
	1100	589.2	5.359	29(2)
	1200	590.9	5.345	28(1)
	1300	589.8	5.354	33(2)
	1400	591.2	5.341	34(3)
Ni$_{0.8}$Zn$_{0.2}$Fe$_2$O$_4$	900	585.3	5.395	30(2)
Mn$_{0.4}$Zn$_{0.6}$Fe$_2$O$_4$	900	600.9	5.236	25(1)
Co$_{0.4}$Zn$_{0.6}$Fe$_2$O$_4$	900	573.3	5.525	21(2)
Ni$_{0.53}$Cu$_{0.12}$Zn$_{0.35}$Fe$_2$O$_4$	900	593.8	5.309	29(6)
Mg$_{0.2}$Cu$_{0.3}$Zn$_{0.5}$Fe$_2$O$_4$	900	596.8	5.171	34(3)
NiFe$_2$O$_4$	700	582.4	5.346	9(1)

V = Unit cell volume
D_x = X-ray density
t = Average grain size

Table 2.9 Structural parameters obtained from the distribution of cations.

Sample	Temp (°C)	Structural parameters from the knowledge of cation distribution		
		a_{the} (Å)	$r_{A(CD)}$ (Å)	$r_{B(CD)}$ (Å)
Ni$_{0.5}$Zn$_{0.5}$Fe$_2$O$_4$	700	8.344	0.640	0.678
	800	8.346	0.650	0.672
	900	8.347	0.652	0.672
	1000	8.348	0.659	0.668
	1100	8.345	0.642	0.676
	1200	8.348	0.658	0.668
	1300	8.349	0.663	0.666
	1400	8.349	0.662	0.667
Ni$_{0.8}$Zn$_{0.2}$Fe$_2$O$_4$	900	8.324	0.640	0.670
Mn$_{0.4}$Zn$_{0.6}$Fe$_2$O$_4$	900	8.435	0.712	0.670
Co$_{0.4}$Zn$_{0.6}$Fe$_2$O$_4$	900	8.373	0.668	0.672
Ni$_{0.53}$Cu$_{0.12}$Zn$_{0.35}$Fe$_2$O$_4$	900	8.349	0.688	0.652
Mg$_{0.2}$Cu$_{0.3}$Zn$_{0.5}$Fe$_2$O$_4$	900	8.361	0.703	0.648
NiFe$_2$O$_4$	700	8.312	0.643	0.664

a_{the} = theoretical lattice parameter
r_A and r_B = tetrahedral and octahedral site radii obtained from cation distributions

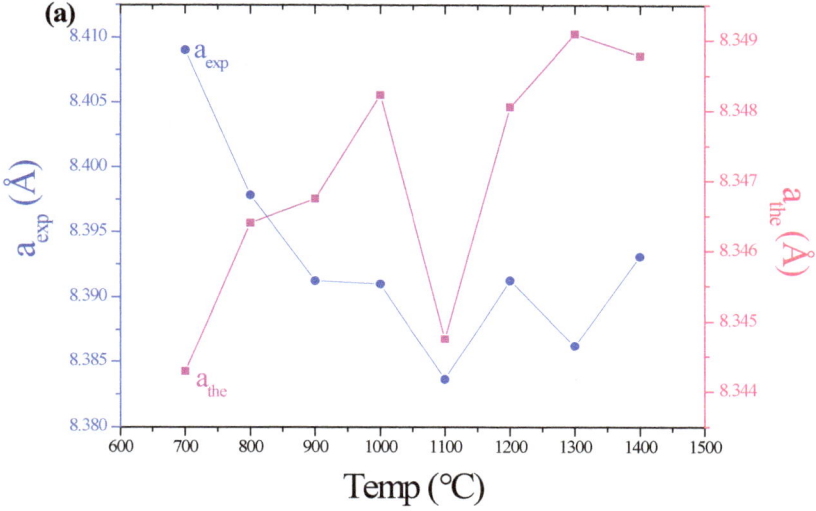

Fig.2.5(a) *Trend in the experimental (a_{exp}) and theoretical (a_{the}) lattice parameter of $Ni_{0.5}Zn_{0.5}Fe_2O_4$ samples.*

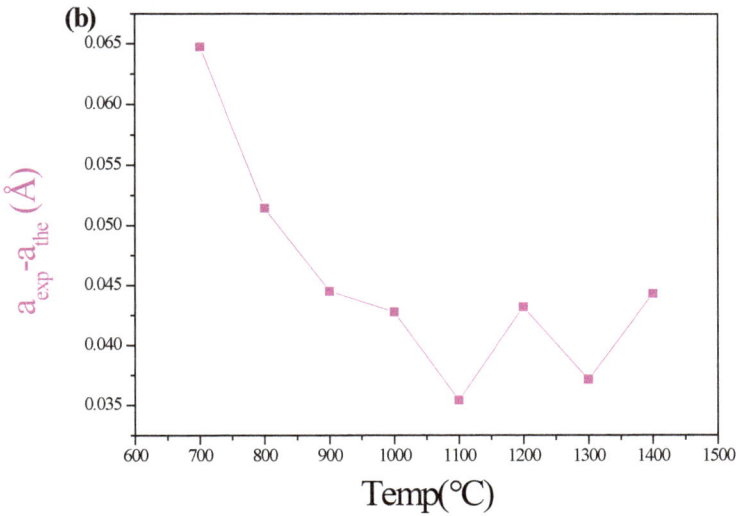

Fig.2.5(b) *Difference between the experimental (a_{exp}) and theoretical (a_{the}) lattice parameter of $Ni_{0.5}Zn_{0.5}Fe_2O_4$ samples.*

2.5 TG/DTA analysis of $Ni_{0.5}Zn_{0.5}Fe_2O_4$ (700 - 1000°C) and $Ni_{0.8}Zn_{0.2}Fe_2O_4$ samples

TG/DTA analysis was carried out on the samples $Ni_{0.5}Zn_{0.5}Fe_2O_4$ sintered from 700-1000°C prepared by mechanical alloying method and $Ni_{0.8}Zn_{0.2}Fe_2O_4$ sintered at 900°C. TG/DTA analyses were carried out on the "as prepared" samples to find the optimum calcinations temperature. But in this present research work, TG/DTA analyses were carried out on the samples after final sintering, to study the evolution of a second phase. The TG/DTA diagrams of the samples $Ni_{0.5}Zn_{0.5}Fe_2O_4$ (700-1000°C) and $Ni_{0.8}Zn_{0.2}Fe_2O_4$ are shown in Figs.2.6(a-e).

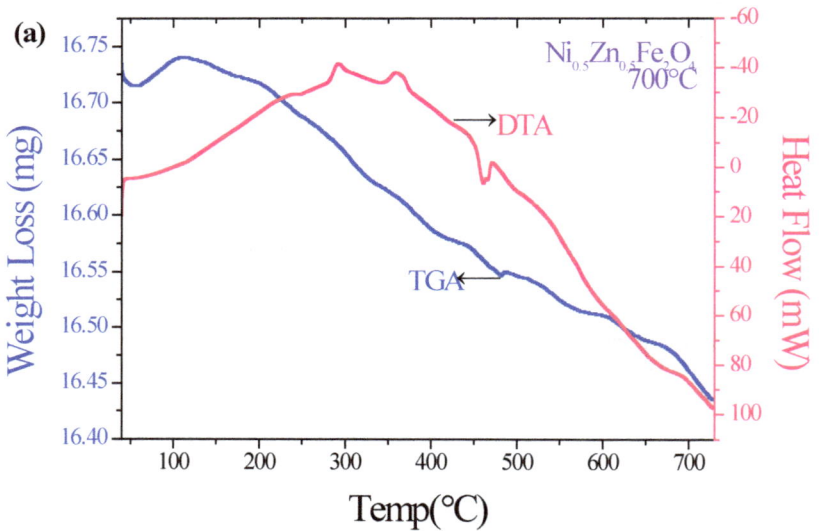

Fig.2.6(a) TG-DTA of $Ni_{0.5}Zn_{0.5}Fe_2O_4$ sample sintered at 700°C.

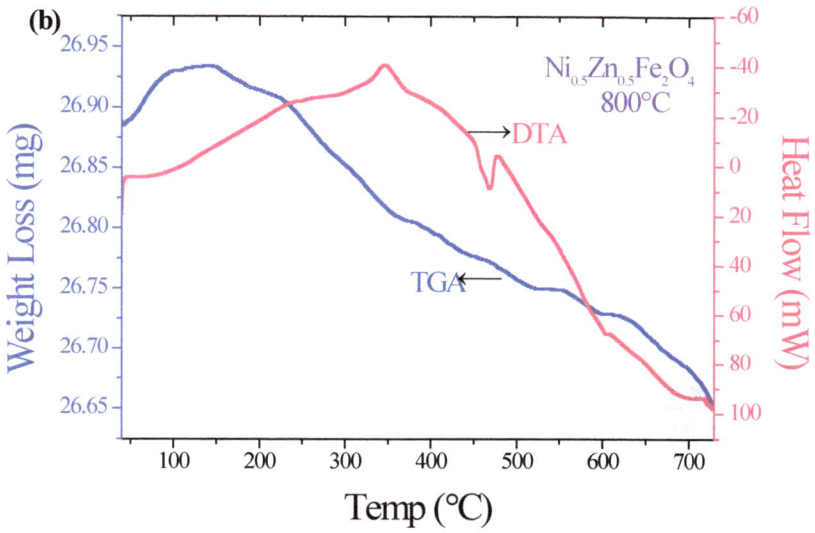

Fig.2.6(b) TG-DTA of $Ni_{0.5}Zn_{0.5}Fe_2O_4$ sample sintered at 800°C.

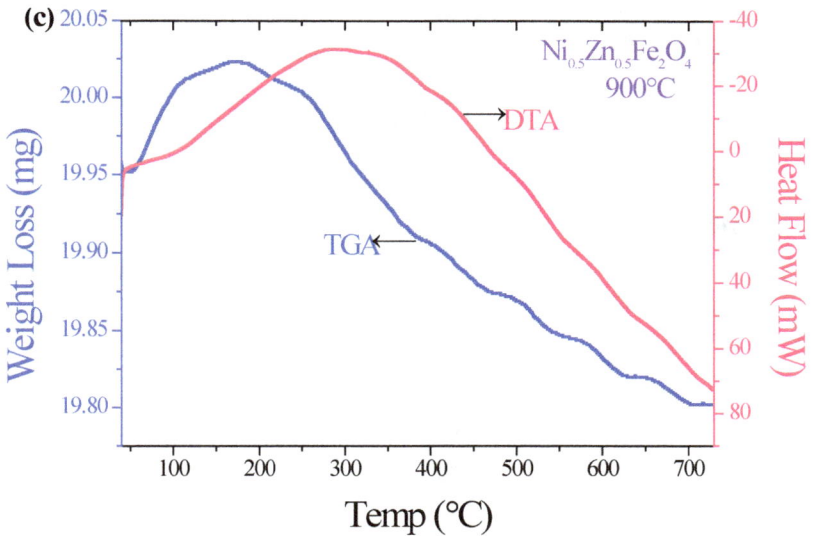

Fig.2.6(c) TG-DTA of $Ni_{0.5}Zn_{0.5}Fe_2O_4$ sample sintered at 900°.

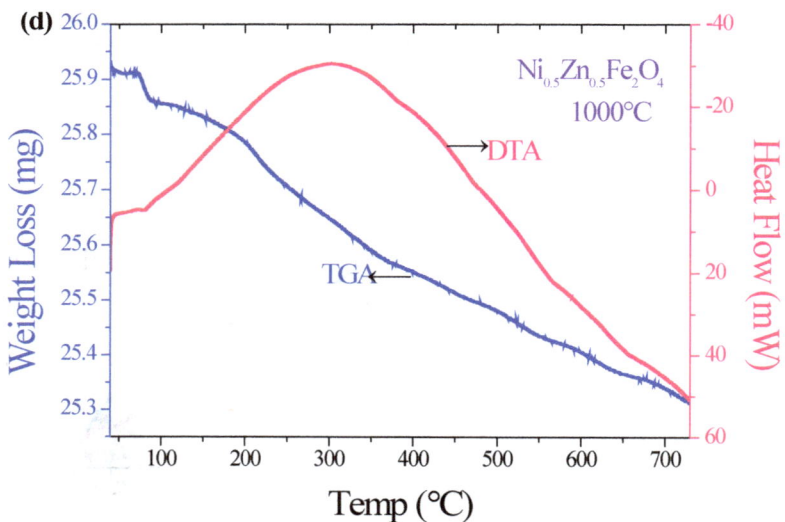

Fig.2.6(d) TG-DTA of $Ni_{0.5}Zn_{0.5}Fe_2O_4$ sample sintered at 1000°C.

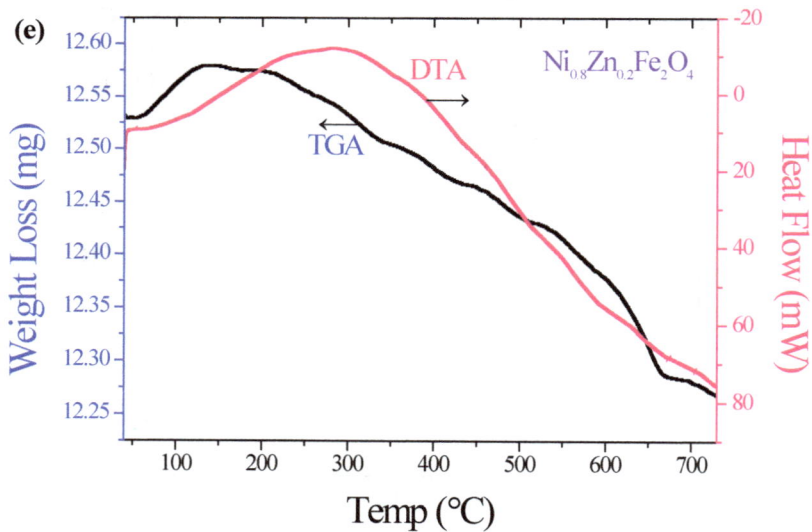

Fig.2.6(e) TG-DTA of $Ni_{0.8}Zn_{0.2}Fe_2O_4$ sample sintered at 900°C.

2.6   Morphological studies of synthesized samples

2.6.1   SEM results

The scanning electron microscope (SEM) images can be used to analyze the presence of porosity, the effect of sintering temperature on the particle size (in the case of $Ni_{0.5}Zn_{0.5}Fe_2O_4$ 700-1400°C samples), nature of distribution and size of the particles in the samples synthesized in the present research work. The dielectric and conducting property of the samples can be correlated with the porosity and grain size of the particles (Beh et al., 2014). The SEM images of all the synthesized samples are shown in Figs.2.7(a-n). Table 2.10 lists the particle size evaluated from SEM images and the grain size evaluated from the software GRAIN (Saravanan R, GRAIN software).

Fig.2.7(a) *SEM image of $Ni_{0.5}Zn_{0.5}Fe_2O_4$ sample sintered at 700°C.*

(b) $Ni_{0.5}Zn_{0.5}Fe_2O_4$ - 800°C

Fig.2.7(b) SEM image of $Ni_{0.5}Zn_{0.5}Fe_2O_4$ sample sintered at 800°C.

(c) $Ni_{0.5}Zn_{0.5}Fe_2O_4$ - 900°C

Fig.2.7(c) SEM image of $Ni_{0.5}Zn_{0.5}Fe_2O_4$ sample sintered at 900°C.

(d) $Ni_{0.5}Zn_{0.5}Fe_2O_4$ - 1000°C

Fig.2.7(d) SEM image of $Ni_{0.5}Zn_{0.5}Fe_2O_4$ sample sintered at 1000°C.

(e) $Ni_{0.5}Zn_{0.5}Fe_2O_4$ - 1100°C

Fig.2.7(e) SEM image of $Ni_{0.5}Zn_{0.5}Fe_2O_4$ sample sintered at 1100°C.

69

(f) $Ni_{0.5}Zn_{0.5}Fe_2O_4$ - 1200°C

Fig.2.7(f) SEM image of $Ni_{0.5}Zn_{0.5}Fe_2O_4$ sample sintered at 1200°C.

(g) $Ni_{0.5}Zn_{0.5}Fe_2O_4$ - 1300°C

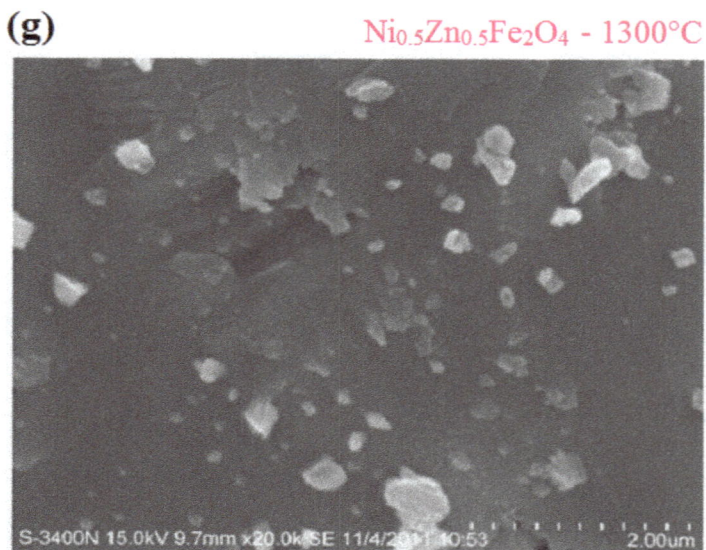

Fig.2.7(g) SEM image of $Ni_{0.5}Zn_{0.5}Fe_2O_4$ sample sintered at 1300°C.

(h) $Ni_{0.5}Zn_{0.5}Fe_2O_4 - 1400°C$

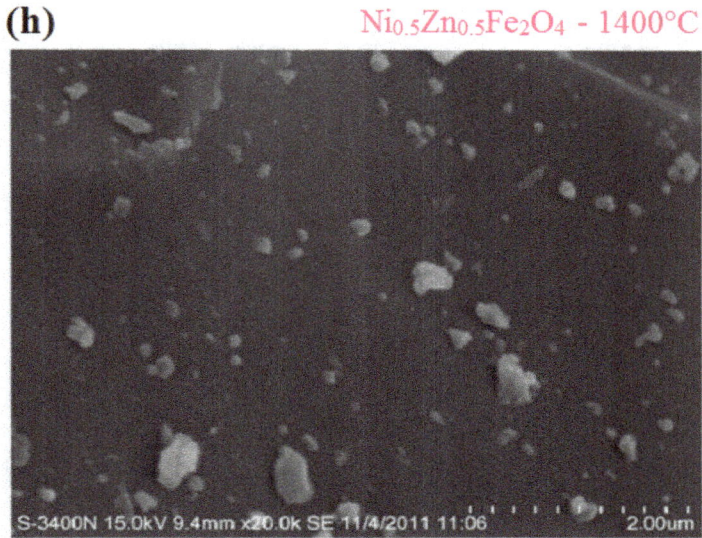

Fig.2.7(h) SEM image of $Ni_{0.5}Zn_{0.5}Fe_2O_4$ sample sintered at 1400°C.

(i) $Ni_{0.8}Zn_{0.2}Fe_2O_4$

Fig.2.7(i) SEM image of $Ni_{0.8}Zn_{0.2}Fe_2O_4$ sample.

(j)

$Mn_{0.4}Zn_{0.6}Fe_2O_4$

S-3400N 15.0kV 9.6mm x20.0k SE 11/3/2011 10:43   2.00um

Fig.2.7(j) *SEM image of* $Mn_{0.4}Zn_{0.6}Fe_2O_4$ *sample.*

(k)

$Co_{0.4}Zn_{0.6}Fe_2O_4$

S-3400N 15.0kV 9.7mm x20.0k SE 11/3/2011 10:57   2.00um

Fig.2.7(k) *SEM image of* $Co_{0.4}Zn_{0.6}Fe_2O_4$ *sample.*

(l) $Ni_{0.53}Cu_{0.12}Zn_{0.35}Fe_2O_4$

Fig.2.7(l) SEM image of $Ni_{0.53}Cu_{0.12}Zn_{0.35}Fe_2O_4$ sample.

(m) $Mg_{0.2}Cu_{0.3}Zn_{0.5}Fe_2O_4$

Fig.2.7(m) SEM image of $Mg_{0.2}Cu_{0.3}Zn_{0.5}Fe_2O_4$ sample.

(n) $NiFe_2O_4$

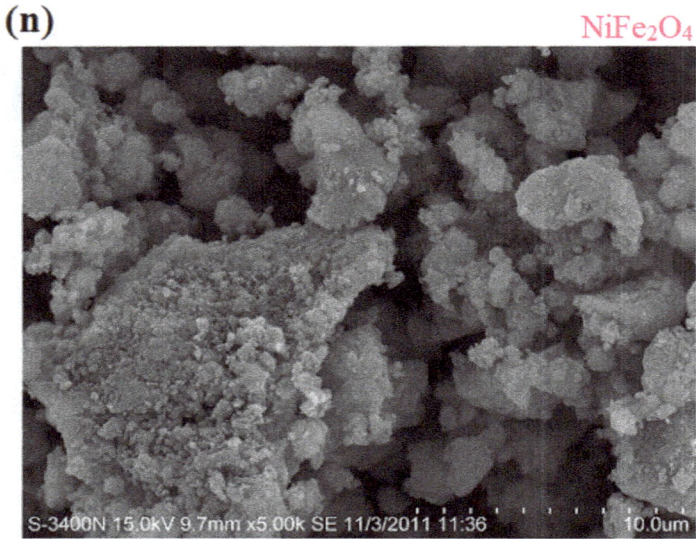

S-3400N 15.0kV 9.7mm x5.00k SE 11/3/2011 11:36   10.0um

Fig.2.7(n) *SEM image of $NiFe_2O_4$ sample.*

Table 2.10 *Estimation of particle size of the synthesized samples from SEM pictures and grain size from powder XRD.*

Sample	Temp °C	Particle size from SEM (µm)	Grain size from XRD (µm)	N*
	700	0.18	0.0081	22
	800	0.19	0.0086	22
	900	0.22	0.0087	25
$Ni_{0.5}Zn_{0.5}Fe_2O_4$	1000	0.63	0.0088	71
	1100	0.97	0.0140	69
	1200	1.24	0.0124	100
	1300	33.0	0.0132	2500
	1400	25.6	0.0128	2000
$Ni_{0.8}Zn_{0.2}Fe_2O_4$	900	0.24	0.0111	21
$Mn_{0.4}Zn_{0.6}Fe_2O_4$	900	1.10	0.0102	107
$Co_{0.4}Zn_{0.6}Fe_2O_4$	900	0.44	0.0234	19
$Ni_{0.53}Cu_{0.12}Zn_{0.35}Fe_2O_4$	900	0.45	0.0101	44
$Mg_{0.2}Cu_{0.3}Zn_{0.5}Fe_2O_4$	900	0.86	0.0106	81
$NiFe_2O_4$	700	0.82	0.0081	101

*N = Particle size from SEM / Grain size from XRD.

2.6.2 EDAX and XRF characterization results

EDAX and XRF results were utilized to assess the concentration of various elements present, stoichiometric ratios and the presence of any impurities in the sample (Hankare et al., 2013) EDAX spectrum and elemental composition of elements for $Ni_{0.5}Zn_{0.5}Fe_2O_4$ samples sintered from $900 - 1200°C$, $Ni_{0.8}Zn_{0.2}Fe_2O_4$, $Mn_{0.4}Zn_{0.6}Fe_2O_4$, and $NiFe_2O_4$ samples are shown in Figs.2.8(a-g) respectively. Tables 2.11(a-d) tabulates the expected stiochiometry and estimated stiochiometry, which is obtained from EDAX result. XRF results of all samples synthesized in this research work are tabulated in Tables 2.12(a-e).

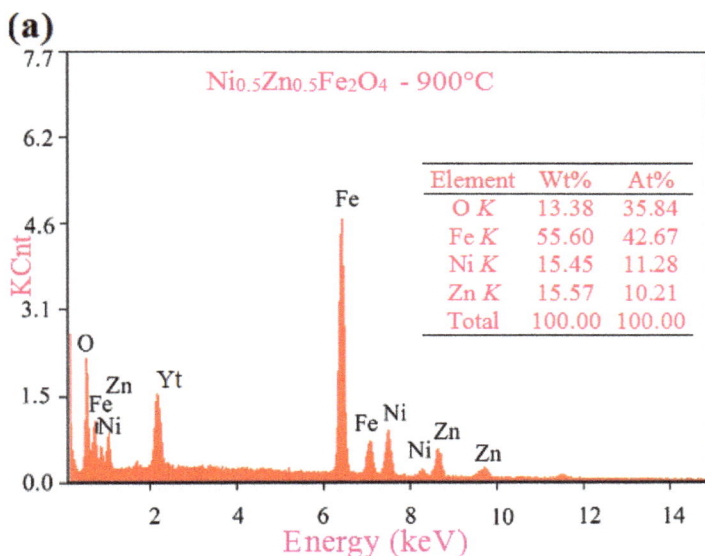

Fig.2.8(a) EDAX spectrum of $Ni_{0.5}Zn_{0.5}Fe_2O_4$ sample sintered at 900°C.

(b)

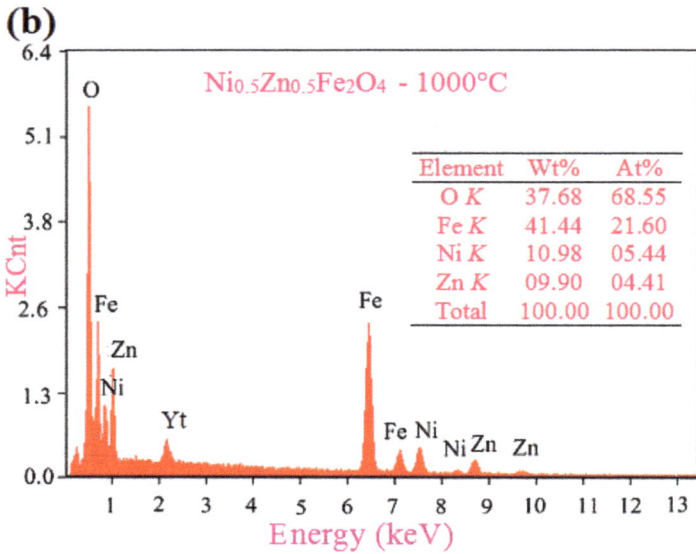

Element	Wt%	At%
O K	37.68	68.55
Fe K	41.44	21.60
Ni K	10.98	05.44
Zn K	09.90	04.41
Total	100.00	100.00

Fig.2.8(b) EDAX spectrum of $Ni_{0.5}Zn_{0.5}Fe_2O_4$ sample sintered at 1000°C.

(c)

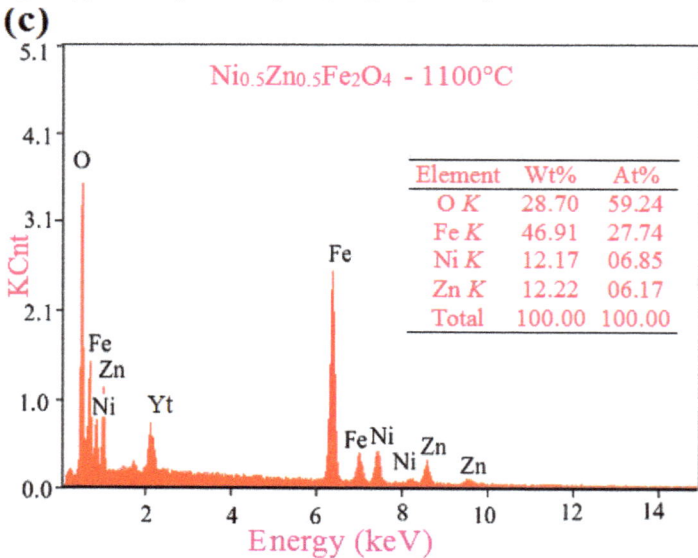

Element	Wt%	At%
O K	28.70	59.24
Fe K	46.91	27.74
Ni K	12.17	06.85
Zn K	12.22	06.17
Total	100.00	100.00

Fig.2.8(c) EDAX spectrum of $Ni_{0.5}Zn_{0.5}Fe_2O_4$ sample sintered at 1100°C.

(d)

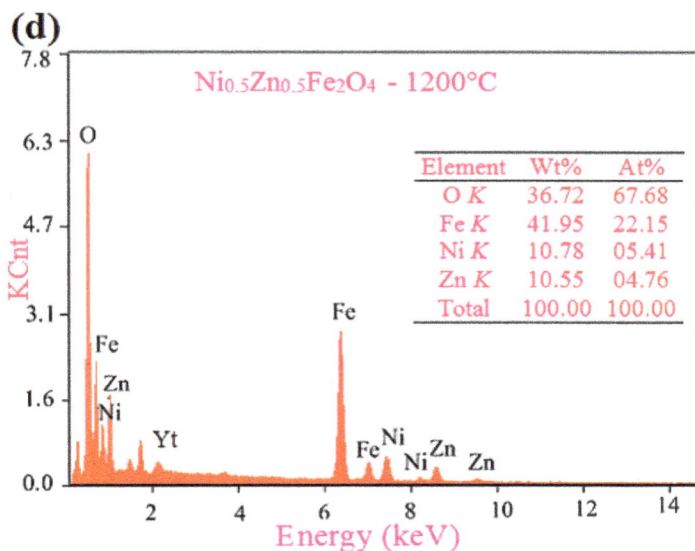

Fig.2.8(d) *EDAX spectrum of* $Ni_{0.5}Zn_{0.5}Fe_2O_4$ *sample sintered at 1200°C.*

(e)

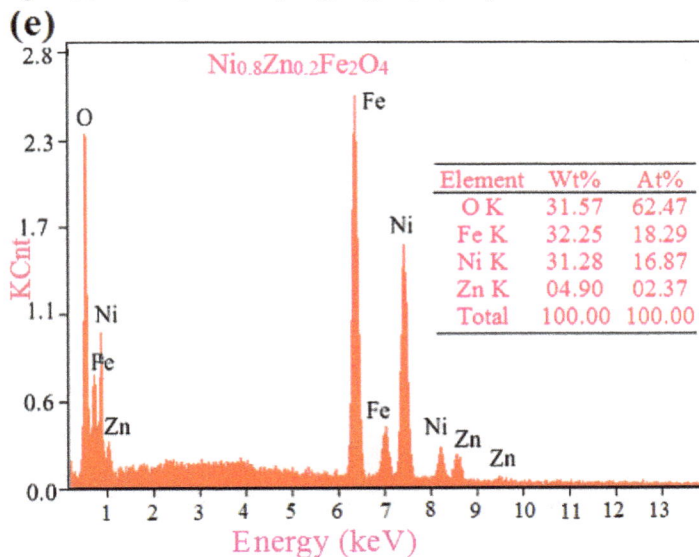

Fig.2.8(e) *EDAX spectrum of* $Ni_{0.8}Zn_{0.2}Fe_2O_4$ *sample.*

77

(f)

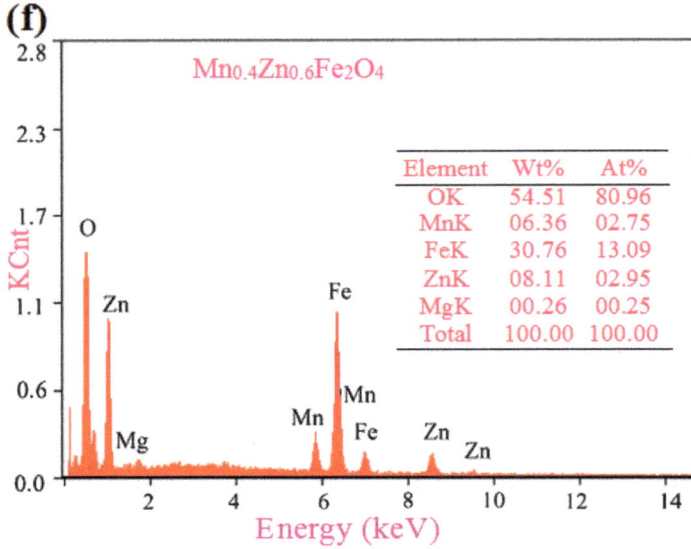

Fig.2.8(f) EDAX spectrum of $Mn_{0.4}Zn_{0.6}Fe_2O_4$ sample.

(g)

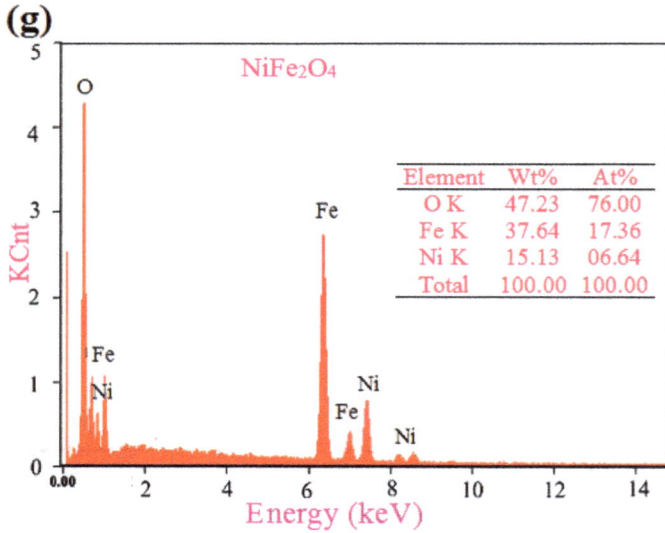

Fig.2.8(g) EDAX spectrum of $NiFe_2O_4$ sample.

Table 2.11(a) *Expected and estimated stiochiometry of $Ni_{0.5}Zn_{0.5}Fe_2O_4$ samples from EDAX spectrum.*

Sample		Expected stiochiometry			Estimated stiochiometry		
		Zn/Fe	Ni/Fe	Ni/Zn	Zn/Fe	Ni/Fe	Ni/Zn
	900°C	0.25	0.25	1.0	0.28	0.28	0.99
$Ni_{0.5}Zn_{0.5}Fe_2O_4$	1000°C	0.25	0.25	1.0	0.27	0.24	1.10
	1100°C	0.25	0.25	1.0	0.26	0.26	0.99
	1200°C	0.25	0.25	1.0	0.25	0.26	1.02

Table 2.11(b) *Expected and estimated stiochiometry of $Ni_{0.8}Zn_{0.2}Fe_2O_4$ samples from EDAX spectrum.*

Sample	Expected stiochiometry			Estimated stiochiometry		
	Zn/Fe	Ni/Fe	Ni/Zn	Zn/Fe	Ni/Fe	Ni/Zn
$Ni_{0.8}Zn_{0.2}Fe_2O_4$	0.1	0.4	4.0	0.15	0.97	6.38

Table 2.11(c) *Expected and estimated stiochiometry of $Mn_{0.4}Zn_{0.6}Fe_2O_4$ samples from EDAX spectrum.*

Sample	Expected stiochiometry			Estimated stiochiometry		
	Mn/Fe	Zn/Fe	Mn/Zn	Mn/Fe	Zn/Fe	Mn/Zn
$Mn_{0.4}Zn_{0.6}Fe_2O_4$	0.2	0.3	0.67	0.21	0.26	0.78

Table 2.11(d) *Expected and estimated stiochiometry of $NiFe_2O_4$ samples from EDAX spectrum.*

Sample	Expected stiochiometry	Estimated stiochiometry
	Ni/Fe	Ni/Fe
$NiFe_2O_4$	0.5	0.4

Table 2.12(a) *Concentration of various elements present in $Ni_{0.5}Zn_{0.5}Fe_2O_4$ sintered at 1400°C samples obtained from XRF analysis.*

$Ni_{0.5}Zn_{0.5}Fe_2O_4$ - 1400°C	
Formula	Concentration (%)
Ni	14.320
Zn	15.250
Fe	42.020
O	26.900

Table 2.12(b) *Concentration of various elements present in $Ni_{0.8}Zn_{0.2}Fe_2O_4$ samples obtained from XRF analysis.*

$Ni_{0.8}Zn_{0.2}Fe_2O_4$	
Element	**Concentration (%)**
Ni	29.010
Zn	06.327
Fe	34.560
O	26.300
Mg	00.254

Table 2.12(c) *Concentration of various elements present in $Mn_{0.4}Zn_{0.6}Fe_2O_4$ and $Co_{0.4}Zn_{0.6}Fe_2O_4$ samples obtained from XRF analysis.*

$Mn_{0.4}Zn_{0.6}Fe_2O_4$		$Co_{0.4}Zn_{0.6}Fe_2O_4$	
Element	**Concentration (%)**	**Element**	**Concentration (%)**
Zn	14.320	Zn	19.620
Fe	43.160	Fe	21.670
O	27.400	O	26.400
Mn	07.695	Co	26.050
Cu	00.027	Cu	00.111
Ni	00.024	Ni	00.061

Table 2.12(d) *Concentration of various elements present in $Ni_{0.53}Cu_{0.12}Zn_{0.35}Fe_2O_4$ and $Mg_{0.2}Cu_{0.3}Zn_{0.5}Fe_2O_4$ samples obtained from XRF analysis.*

$Ni_{0.53}Cu_{0.12}Zn_{0.35}Fe_2O_4$		$Mg_{0.2}Cu_{0.3}Zn_{0.5}Fe_2O_4$	
Element	**Concentration (%)**	**Element**	**Concentration (%)**
Ni	18.750	Mg	01.200
Zn	12.450	Zn	21.670
Fe	33.420	Fe	34.250
O	25.200	O	26.700
Cu	03.896	Cu	12.690
--	---	Ni	00.849

Table 2.12(e) *Concentration of various elements present in $NiFe_2O_4$ samples obtained from XRF analysis.*

$NiFe_2O_4$	
Element	**Concentration (%)**
Ni	31.680
Fe	34.720
O	26.500

2.7   Results – MEM analysis

MEM method is employed to determine the mid bond electron density in the untit cell of ferrites, using the PRIMA software (Izumi and Dilanian, 2002). MEM results are utilized in the VESTA software (Momma and Izumi, 2006) to construct 3D, 2D and 1D electron density maps. Numerical values of the mid bond electron density values can be obtained from 1D electron density maps.

Based on the mid bond electron density values, the strength or weakness of various site interactions (A-A, A-B and B-B) in the unit cell of ferrite was evaluated. Hitherto, the strengthening or weakening of various site interactions was predicted based on the interionic distances (M_e – O, M_e – M_e) and interionic bond angles (θ_1 – θ_5) values. The interionic distances between cation – anion (M_e – O), cation – cation (M_e – M_e) and interionic bond angles (θ_1 – θ_5) are calculated using Eqs.(1.42 – 1.55) for all samples and are tabulated in Tables 2.13(a-c). Refinement parameters obtained from MEM are tabulated in Table 2.14. The 3D isosurfaces, which represent the unit cell of the ferrite, of all samples synthesized in the present study are shown in Figs.2.9(a-n). The 2D contour maps are drawn with contour lines between 0 to 1 e/$Å^3$ with a contour interval of 0.1e/$Å^3$ for all the samples in the present research work, except $Ni_{0.53}Cu_{0.12}Zn_{0.35}Fe_2O_4$, $Mg_{0.2}Cu_{0.3}Zn_{0.2}Fe_2O_4$ in which the contour interval is 0.15e/$Å^3$, are shown in Figs.2.10(a-n). 2D contour maps will give a bird's eye view of various sites interaction in a given hkl plane. The 1D profiles are shown in Figs.2.11-2.15 and mid bond electron density of all the samples synthesized in this present research work is tabulated in Table 2.15.

Table 2.13(a) *Interionic distances and interionic angles of $Ni_{0.5}Zn_{0.5}Fe_2O_4$ samples sintered from 700-1400 °C.*

Inter ionic distances and angles		700°C	800°C	900°C	1000°C
$M_e - M_e$ (Å)	b	2.973	2.969	2.967	2.967
	c	3.486	3.481	3.479	3.479
	d	3.641	3.636	3.634	3.634
	e	5.461	5.454	5.452	5.452
	f	5.149	5.142	5.139	5.139
$M_e - O$ (Å)	p	2.063	2.062	2.073	2.040
	q	1.889	1.886	1.859	1.915
	r	3.617	3.611	3.560	3.667
	s	3.664	3.658	3.647	3.666
Angles(°)	θ_1	123°45′	123°48′	124°19′	123°05′
	θ_2	146°50′	146°52′	149°35′	143°50′
	θ_3	92°12′	92°14′	91°21′	93°14′
	θ_4	125°46′	125°48′	125°34′	126°00′
	θ_5	75°37′	75°38′	77°12′	73°48′
Inter ionic distances and angles		1100°C	1200°C	1300°C	1400°C
$M_e - M_e$ (Å)	b	2.964	2.967	2.965	2.967
	c	3.476	3.479	3.477	3.480
	d	3.630	3.634	3.631	3.634
	e	5.445	5.450	5.447	5.451
	f	5.134	5.139	5.136	5.140
$M_e - O$ (Å)	p	2.020	1.979	2.031	1.969
	q	1.946	2.022	1.930	2.040
	r	3.727	3.872	3.695	3.907
	s	3.674	3.702	3.670	3.709
Angles(°)	θ_1	122°38′	120°78′	122°74′	120°39′
	θ_2	140°88′	134°78′	142°39′	133°40′
	θ_3	94°38′	97°08′	93°78′	97°76′
	θ_4	126°25′	126°82′	126°12′	126°96′
	θ_5	71°94′	67°95′	72°89′	67°04′

Table 2.13(b) *Interionic distances and interionic angles of $Ni_{0.8}Zn_{0.2}Fe_2O_4$, $Mn_{0.4}Zn_{0.6}Fe_2O_4$ and $Co_{0.4}Zn_{0.6}Fe_2O_4$ samples.*

Interionic distances and angles		$Ni_{0.8}Zn_{0.2}Fe_2O_4$	$Mn_{0.4}Zn_{0.6}Fe_2O_4$	$Co_{0.4}Zn_{0.6}Fe_2O_4$
$M_e - M_e$ (Å)	b	2.983	2.937	2.957
	c	3.498	3.444	3.467
	d	3.654	3.597	3.622
	e	5.481	5.395	5.433
	f	5.167	5.087	5.122
$M_e - O$ (Å)	p	1.925	1.997	2.054
	q	2.146	1.936	1.875
	r	4.110	3.708	3.590
	s	3.760	3.643	3.643
Angles (°)	θ_1	118°60′	122°20′	123°84′
	θ_2	127°42′	140°15′	147°24′
	θ_3	101°09′	94°67′	92°08′
	θ_4	127°61′	126°31′	125°74′
	θ_5	63°02′	71°47′	75°85′

Table 2.13(c) *Interionic distances and interionic angles of $Ni_{0.53}Cu_{0.12}Zn_{0.35}Fe_2O_4$, $Mg_{0.2}Cu_{0.3}Zn_{0.5}Fe_2O_4$ and $NiFe_2O_4$ samples.*

Interionic distances and angles		$Ni_{0.53}Cu_{0.12}Zn_{0.35}Fe_2O_4$	$Mg_{0.2}Cu_{0.3}Zn_{0.5}Fe_2O_4$	$NiFe_2O_4$
$M_e - M_e$ (Å)	b	2.972	2.952	2.977
	c	3.485	3.462	3.490
	d	3.640	3.616	3.646
	e	5.459	5.424	5.469
	f	5.147	5.114	5.156
$M_e - O$ (Å)	p	2.041	2.053	2.023
	q	1.925	1.868	1.963
	r	3.686	3.577	3.760
	s	3.674	3.636	3.693
Angles (°)	θ_1	122°95′	123°92′	122°18′
	θ_2	143°26′	147°63′	140°10′
	θ_3	93°46′	91°95′	94°69′
	θ_4	126°05′	125°72′	126°32′
	θ_5	73°44′	76°09′	71°44′

Table 2.14 *MEM refinement parameters of samples synthesized in this research work.*

Sample	Temp (°C)	No. of cycles	F_{000}	F_{000}/V (Å)^3	No.of pixels per unit cell	λ	R_{MEM} (%)	wR_{MEM} (%)
	700	1281	904	1.520	64×64×64	0.0114	0.0305	0.0237
	800	1179	904	1.527	64×64×64	0.0109	0.0227	0.0218
	900	1314	904	1.530	64×64×64	0.0116	0.0299	0.0249
$Ni_{0.5}Zn_{0.5}Fe_2O_4$	1000	1385	904	1.529	64×64×64	0.0119	0.0283	0.0218
	1100	2067	904	1.521	64×64×64	0.0153	0.0306	0.0266
	1200	1352	904	1.532	64×64×64	0.0118	0.0272	0.0214
	1300	2390	904	1.533	64×64×64	0.0169	0.0301	0.0244
	1400	1762	904	1.529	64×64×64	0.0138	0.0286	0.0224
$Ni_{0.8}Zn_{0.2}Fe_2O_4$	900	1469	899	1.536	64×64×64	0.0123	0.0172	0.0167
$Mn_{0.4}Zn_{0.6}Fe_2O_4$	900	2734	896	1.491	64×64×64	0.0187	0.0498	0.0349
$Co_{0.4}Zn_{0.6}Fe_2O_4$	900	2812	902	1.573	64×64×64	0.0191	0.0646	0.0369
$Ni_{0.53}Cu_{0.12}Zn_{0.35}Fe_2O_4$	900	2499	903	1.521	64×64×64	0.0175	0.0235	0.0196
$Mg_{0.2}Cu_{0.3}Zn_{0.5}Fe_2O_4$	900	1423	881	1.476	64×64×64	0.0121	0.0237	0.0207
$NiFe_2O_4$	700	2284	896	1.538	64×64×64	0.0164	0.0529	0.0355

F_{000} = No. of electrons/unit cell
$F_{000}/V(Å)^3$ = uniform prior density
λ = Lagrange parameter
R_{MEM} = reliability index from MEM
wR_{MEM} = weighted reliability index from MEM.

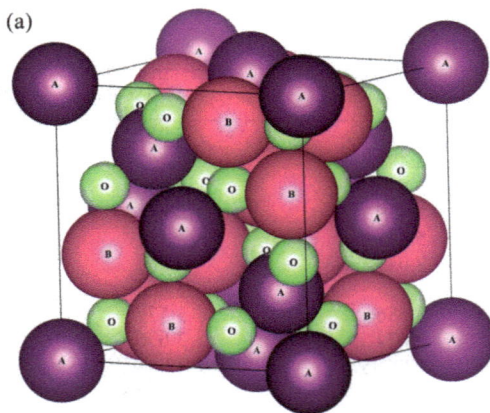

Fig.2.9(a) *Three dimensional electron density view of $Ni_{0.5}Zn_{0.5}Fe_2O_4$ sample sintered at 700°C in space filling style. A, B and O represent tetrahedral sites, octahedral sites and oxygen atoms respectively.*

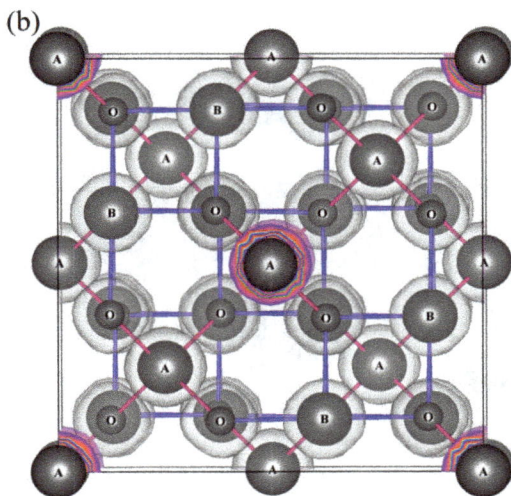

Fig.2.9(b) *Three dimensional electron density view of $Ni_{0.5}Zn_{0.5}Fe_2O_4$ sample sintered at 800°C with isosurface and section volumetric data and also bonds between tetrahedral (A) site – oxygen (O) atoms, octahedral (B) site – oxygen (O) atoms.*

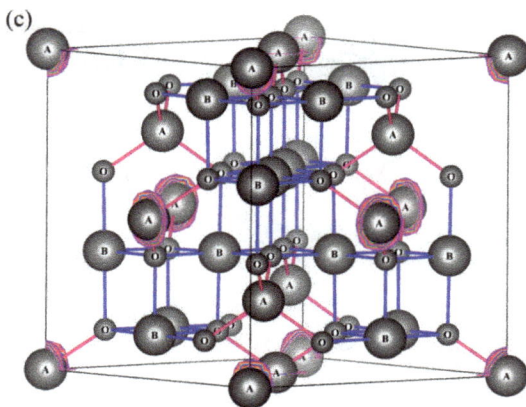

Fig.2.9(c) *Three dimensional electron density view of $Ni_{0.5}Zn_{0.5}Fe_2O_4$ sample sintered at 900°C with section volumetric data and also bonds between tetrahedral (A) sites – oxygen (O) atoms, octahedral (B) sites – oxygen (O) atoms.*

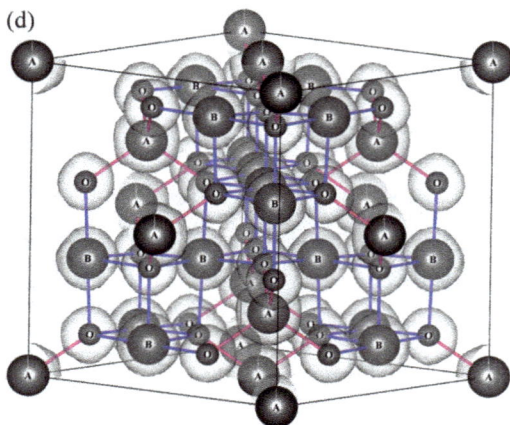

Fig.2.9(d) *Three dimensional electron density view of $Ni_{0.5}Zn_{0.5}Fe_2O_4$ sample sintered at 1000°C with isosurface volumetric data and also bonds between tetrahedral (A) sites – oxygen (O) atoms, octahedral (B) sites – oxygen (O) atoms.*

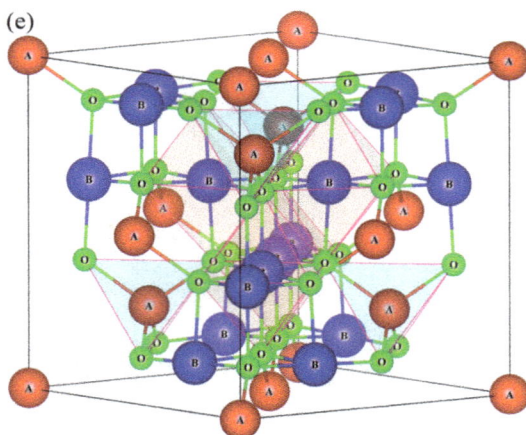

Fig.2.9(e) *Three dimensional electron density view of $Ni_{0.5}Zn_{0.5}Fe_2O_4$ sample sintered at 1100°C with tetrahedral (A) and octahedral (B) site and also bonds between tetrahedral (A) sites – oxygen (O) atoms, octahedral (B) sites – oxygen (O) atoms.*

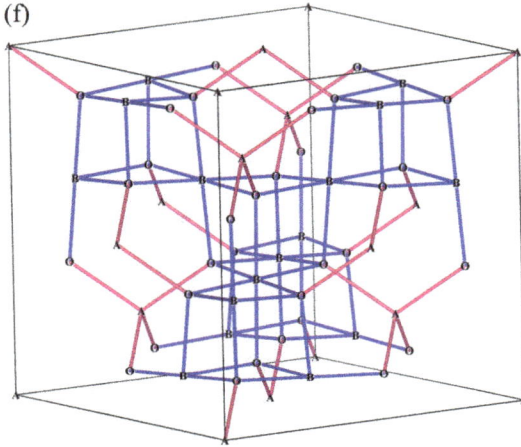

Fig.2.9(f) *Three dimensional electron density view of $Ni_{0.5}Zn_{0.5}Fe_2O_4$ sample sintered at 1200°C with wireframe style and bonds between tetrahedral (A) sites – oxygen (O) atoms, octahedral (B) sites – oxygen (O) atoms.*

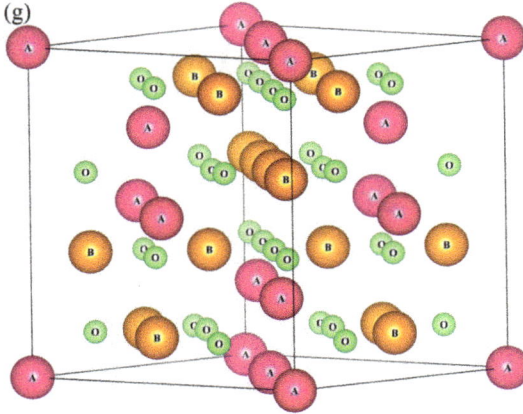

Fig.2.9(g) *Three dimensional electron density view of $Ni_{0.5}Zn_{0.5}Fe_2O_4$ sample sintered at 1300°C with tetrahedral (A) sites, octahedral (B) sites and oxygen atoms.*

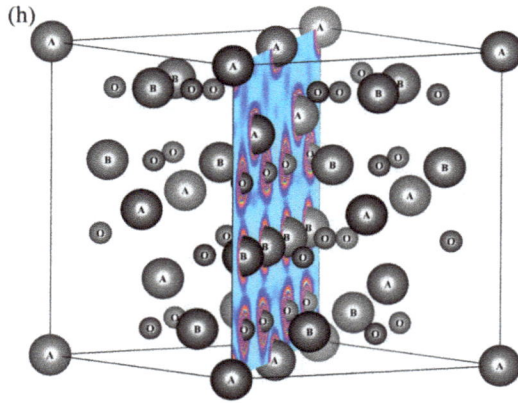

Fig.2.9(h) *Three dimensional electron density view of $Ni_{0.5}Zn_{0.5}Fe_2O_4$ sample sintered at 1400°C with tetrahedral (A) sites, octahedral (B) sites, oxygen atoms and (110) miller indices plane.*

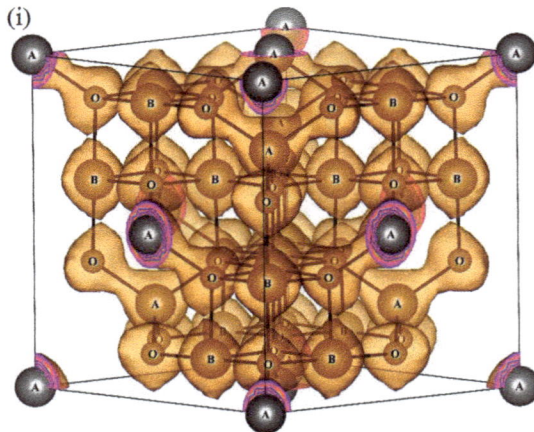

Fig.2.9(i) *Three dimensional electron density view of $Ni_{0.8}Zn_{0.2}Fe_2O_4$ sample with isosurface, section volumetric data and also bonds between tetrahedral (A) sites – oxygen (O) atoms, octahedral (B) sites – oxygen (O) atoms.*

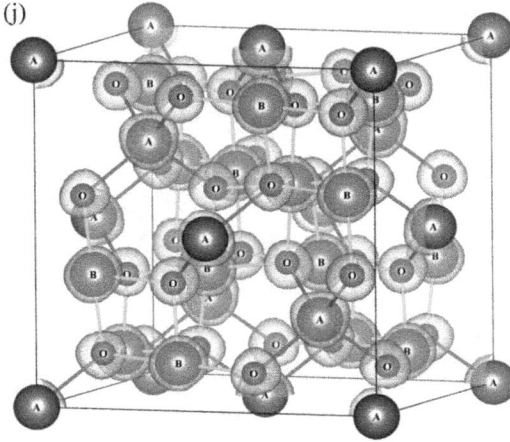

Fig.2.9(j) *Three dimensional electron density view of $Mn_{0.4}Zn_{0.6}Fe_2O_4$ sample with isosurface volumetric data and also bonds between tetrahedral (A) sites – oxygen (O) atoms, octahedral (B) sites – oxygen (O) atoms.*

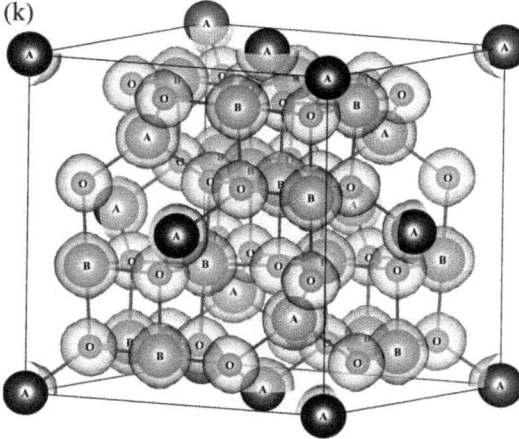

Fig.2.9(k) *Three dimensional electron density view of $Co_{0.4}Zn_{0.6}Fe_2O_4$ sample with isosurface volumetric data and also bonds between tetrahedral (A) sites – oxygen (O) atoms, octahedral (B) sites – oxygen (O) atoms.*

(l)

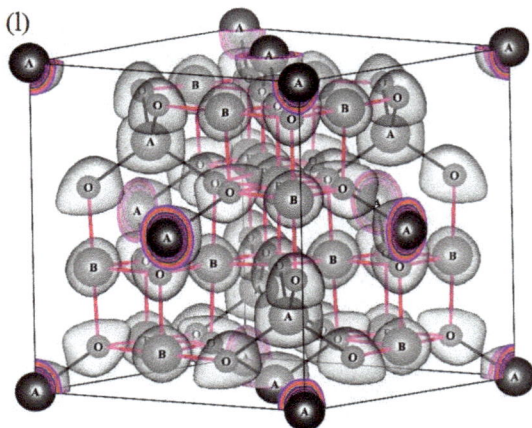

Fig.2.9(l) *Three dimensional electron density view of $Ni_{0.53}Cu_{0.12}Zn_{0.35}Fe_2O_4$ sample with isosurface, section volumetric data and also bonds between tetrahedral (A) sites – oxygen (O) atoms, octahedral (B) sites – oxygen (O) atoms.*

(m)

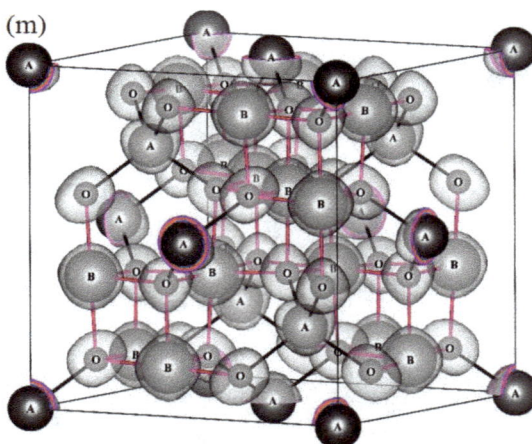

Fig.2.9(m) *Three dimensional electron density view of $Mg_{0.2}Cu_{0.3}Zn_{0.5}Fe_2O_4$ sample with isosurface volumetric data and also bonds between tetrahedral (A) sites – oxygen (O) atoms, octahedral (B) sites – oxygen (O) atoms.*

(n)

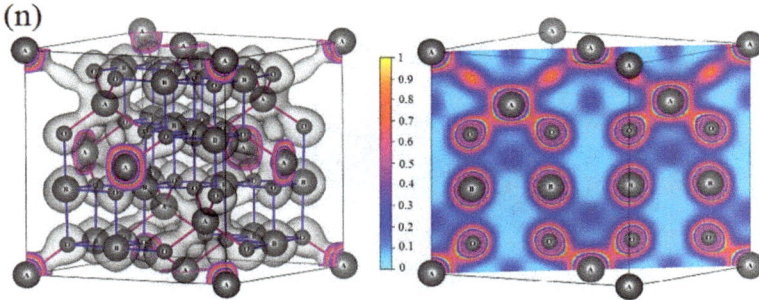

Fig.2.9(n) *Three dimensional (a) electron density view along with two dimensional (b) view of $NiFe_2O_4$ sample with isosurface, section volumetric data and also bonds between tetrahedral (A) sites – oxygen (O) atoms, octahedral (B) sites – oxygen (O) atoms.*

(a)

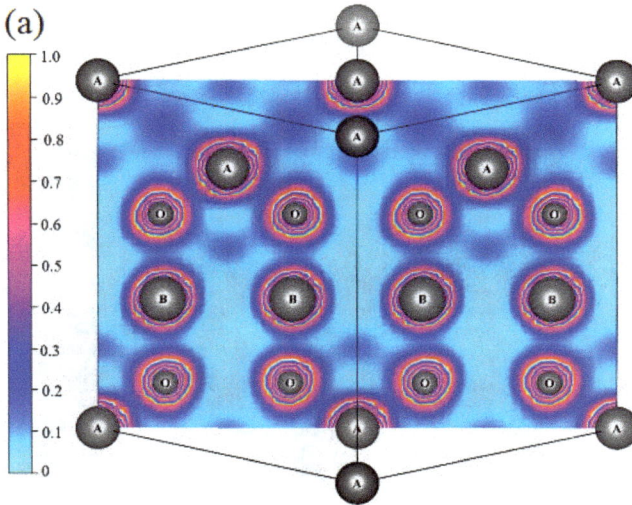

Fig.2.10(a) *2D contour map of $Ni_{0.5}Zn_{0.5}Fe_2O_4$ sintered at 700°C with contour lines between 0 to 1 $e/Å^3$ with a contour interval of 0.1 $e/Å^3$.*

(b)

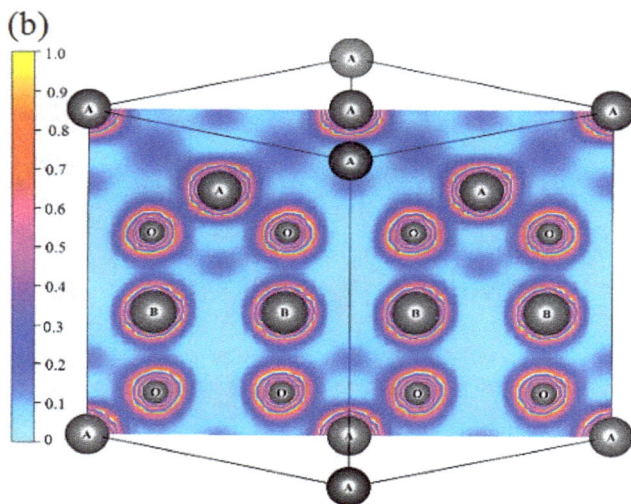

Fig.2.10(b) 2D contour map of $Ni_{0.5}Zn_{0.5}Fe_2O_4$ sintered at 800°C with contour lines between 0 to 1 e/$Å^3$ with a contour interval of 0.1 e/$Å^3$.

(c)

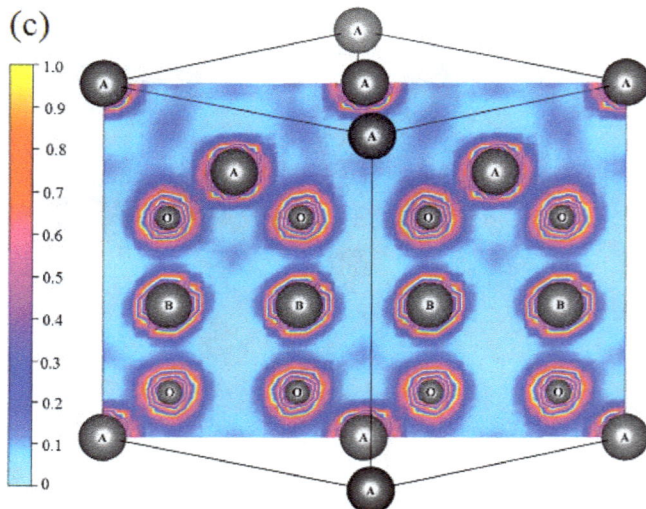

Fig.2.10(c) 2D contour map of $Ni_{0.5}Zn_{0.5}Fe_2O_4$ sintered at 900°C with contour lines between 0 to 1 e/$Å^3$ with a contour interval of 0.1 e/$Å^3$.

(d)

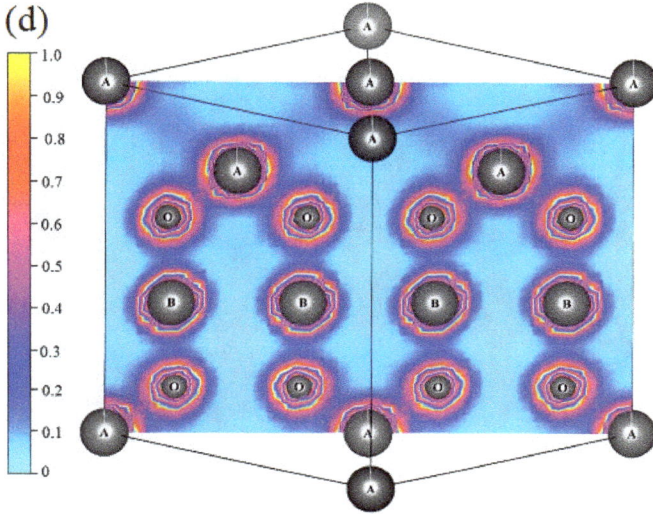

Fig.2.10(d) *2D contour map of $Ni_{0.5}Zn_{0.5}Fe_2O_4$ sintered at 1000°C with contour lines between 0 to 1 e/$Å^3$ with a contour interval of 0.1 e/$Å^3$.*

(e)

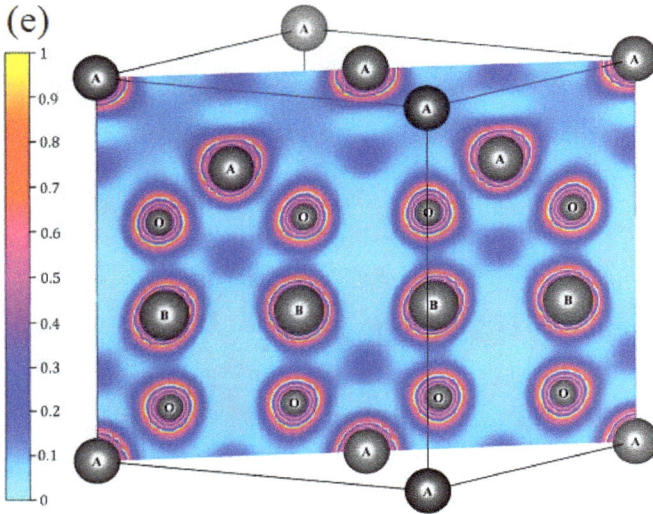

Fig.2.10(e) *2D contour map of $Ni_{0.5}Zn_{0.5}Fe_2O_4$ sintered at 1100°C with contour lines between 0 to 1 e/$Å^3$ with a contour interval of 0.1 e/$Å^3$.*

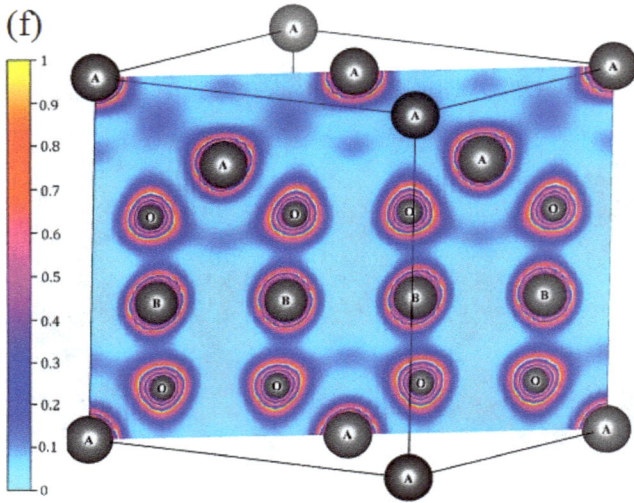

Fig.2.10(f) *2D contour map of $Ni_{0.5}Zn_{0.5}Fe_2O_4$ sintered at 1200°C with contour lines between 0 to 1 e/$Å^3$ with a contour interval of 0.1 e/$Å^3$.*

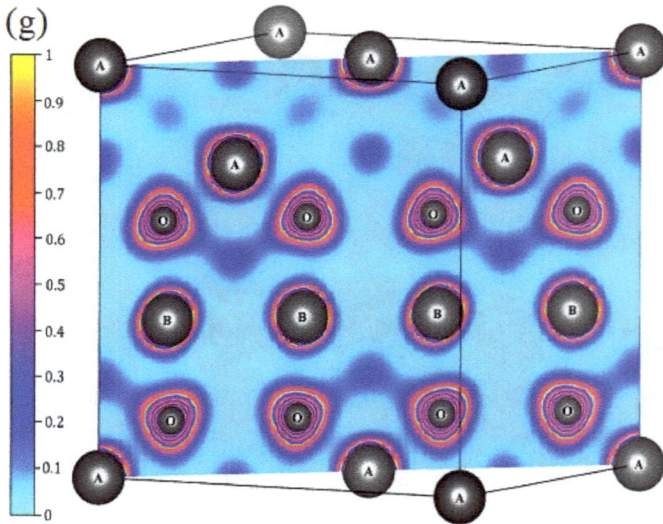

Fig.2.10(g) *2D contour map of $Ni_{0.5}Zn_{0.5}Fe_2O_4$ sintered at 1300°C with contour lines between 0 to 1 e/$Å^3$ with a contour interval of 0.1 e/$Å^3$.*

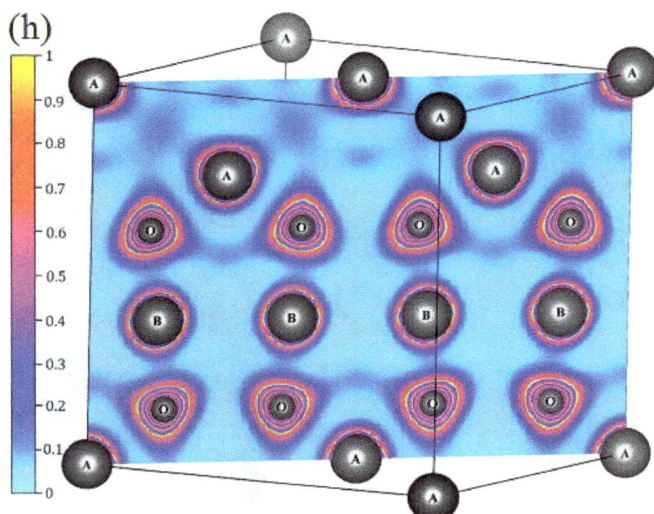

Fig.2.10(h) *2D contour map of $Ni_{0.5}Zn_{0.5}Fe_2O_4$ sintered at 1400°C with contour lines between 0 to 1 e/$\mathring{A}^3$ with a contour interval of 0.1 e/$\mathring{A}^3$.*

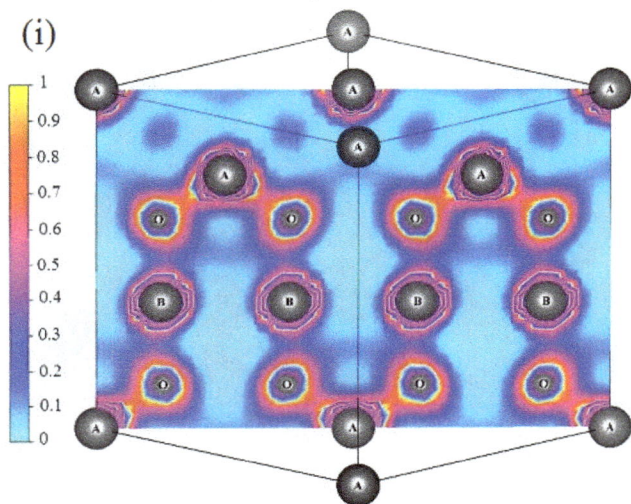

Fig.2.10(i) *2D contour map of $Ni_{0.8}Zn_{0.2}Fe_2O_4$ sample with contour lines between 0 to 1 e/$\mathring{A}^3$ with a contour interval of 0.1 e/$\mathring{A}^3$.*

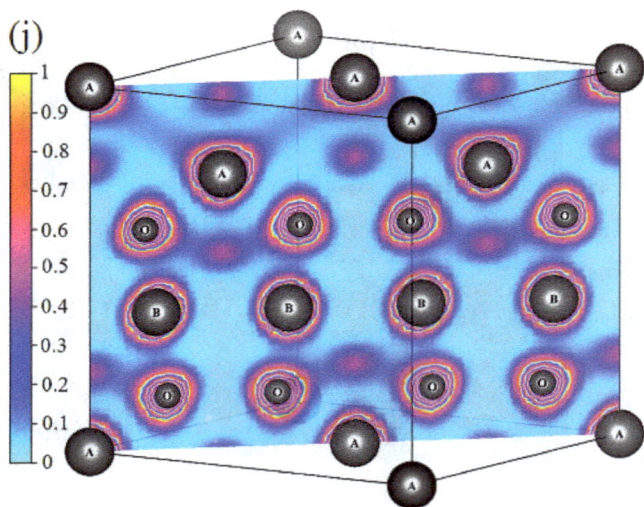

Fig.2.10(j) *2D contour map of $Mn_{0.4}Zn_{0.6}Fe_2O_4$ sample with contour lines between 0 to 1 e/$Å^3$ with a contour interval of 0.1 e/$Å^3$.*

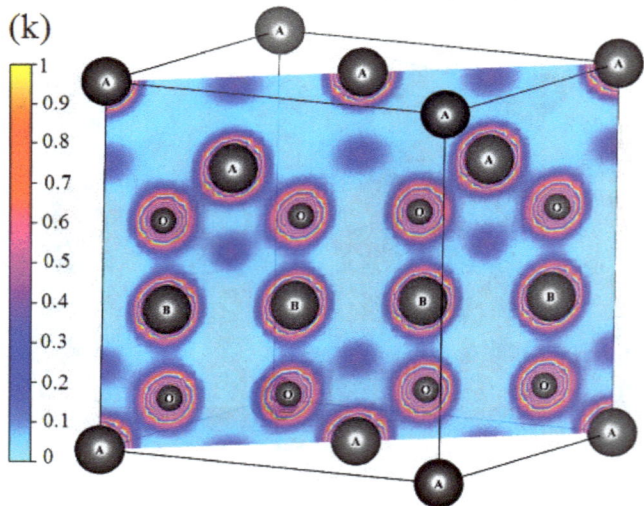

Fig.2.10(k) *2D contour map of $Co_{0.4}Zn_{0.6}Fe_2O_4$ sample with contour lines between 0 to 1 e/$Å^3$ with a contour interval of 0.1 e/$Å^3$.*

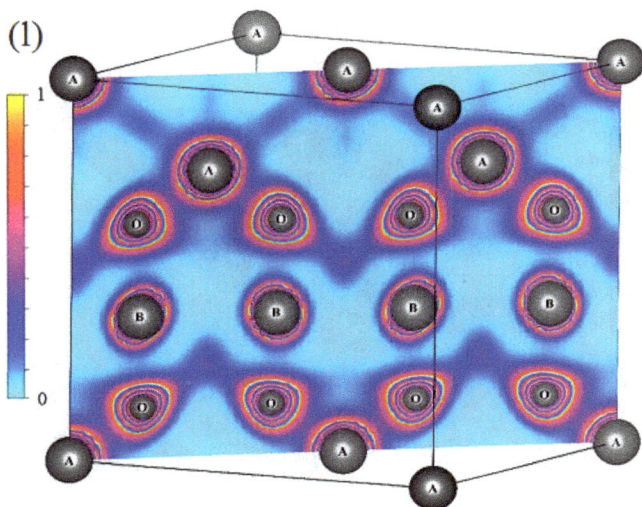

Fig.2.10(l) 2D contour map of $Ni_{0.53}Cu_{0.12}Zn_{0.35}Fe_2O_4$ sample with contour lines between 0 to 1 $e/Å^3$ with a contour interval of 0.15 $e/Å^3$.

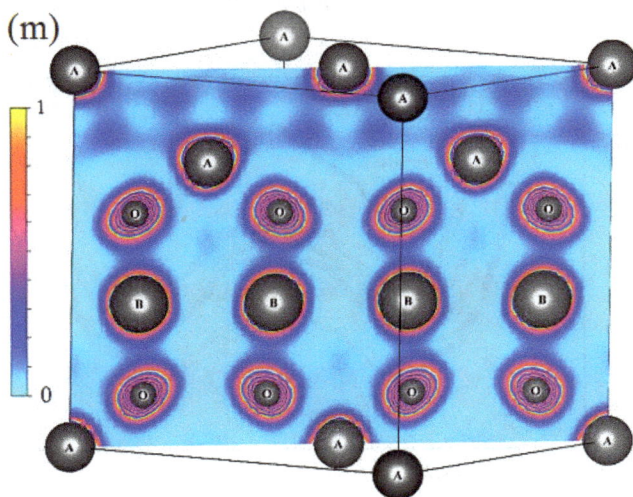

Fig.2.10(m) 2D contour map of $Mg_{0.2}Cu_{0.3}Zn_{0.5}Fe_2O_4$ sample with contour lines between 0 to 1 $e/Å^3$ with a contour interval of 0.15 $e/Å^3$.

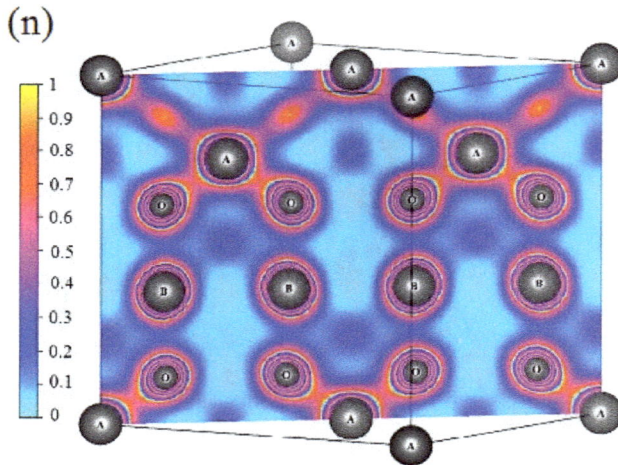

Fig.2.10(n) *2D contour map of $NiFe_2O_4$ sample with contour lines between 0 to 1 $e/Å^3$ with a contour interval of $0.1e/Å^3$.*

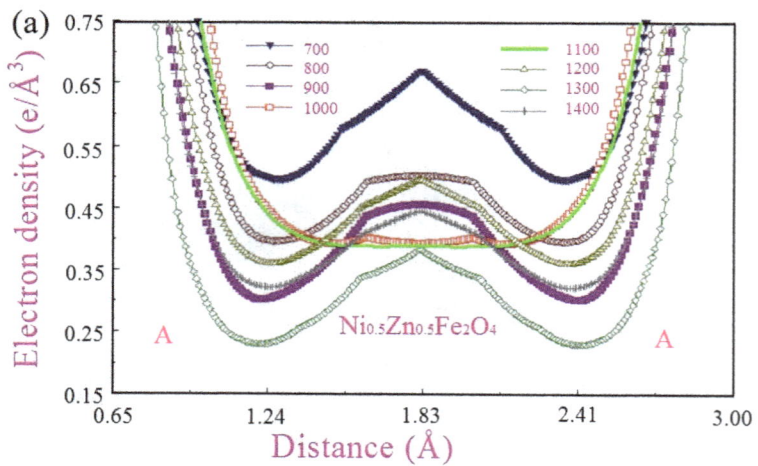

Fig.2.11(a) *One dimensional profile showing mid bond electron density values of tetrahedral (A) - tetrahedral (A) sites interactions of $Ni_{0.5}Zn_{0.5}Fe_2O_4$ nano ferrites samples.*

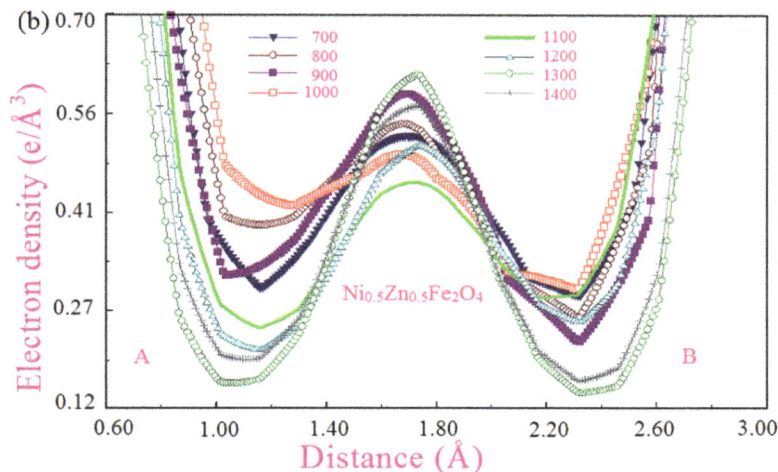

Fig.2.11(b) *One dimensional profile showing mid bond electron density values of tetrahedral (A) site - octahedral (B) site interactions of $Ni_{0.5}Zn_{0.5}Fe_2O_4$ nano ferrites samples.*

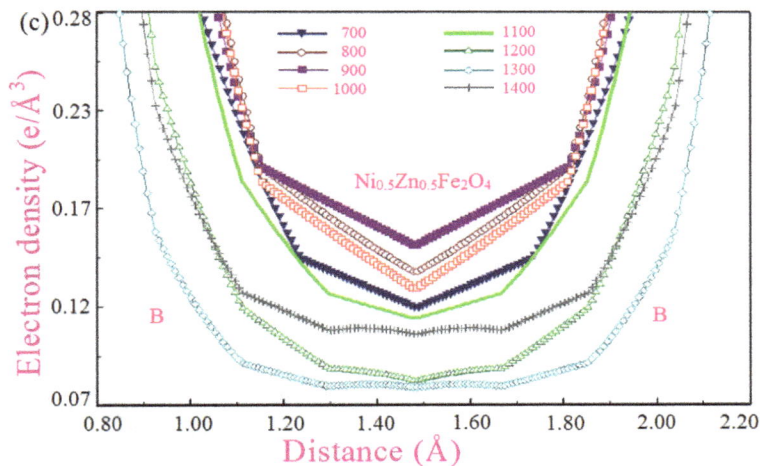

Fig.2.11(c) *One dimensional profile showing mid bond electron density values of octahedral (B) - octahedral (B) sites interactions of $Ni_{0.5}Zn_{0.5}Fe_2O_4$ nano ferrites samples.*

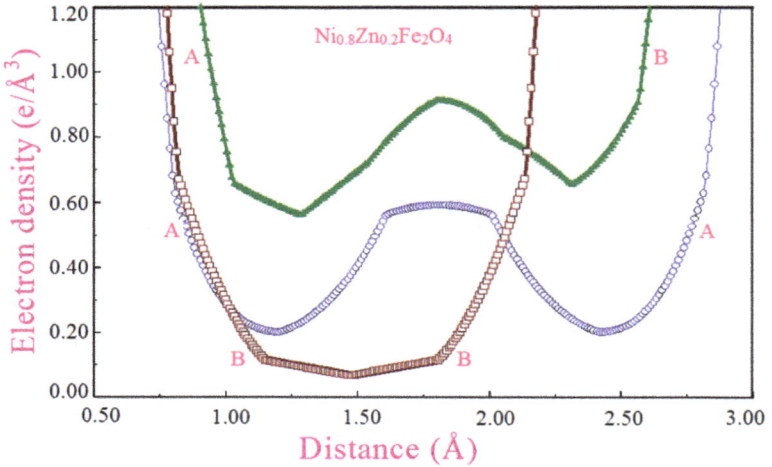

Fig.2.12 *One dimensional profile showing mid bond electron density values of tetrahedral (A) site - tetrahedral (A) site, tetrahedral (A) site - octahedral (B) site, and octahedral (B) site - octahedral (B) site interactions of $Ni_{0.8}Zn_{0.2}Fe_2O_4$ nano ferrites samples.*

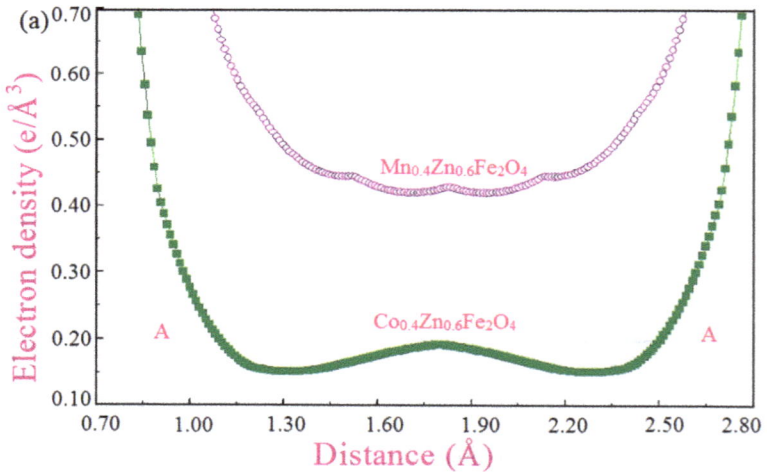

Fig.2.13(a) *One dimensional profile showing mid bond electron density values of tetrahedral (A) site - tetrahedral (A) site interactions of $Mn_{0.4}Zn_{0.6}Fe_2O_4$ and $Co_{0.4}Zn_{0.6}Fe_2O_4$ nano ferrites samples.*

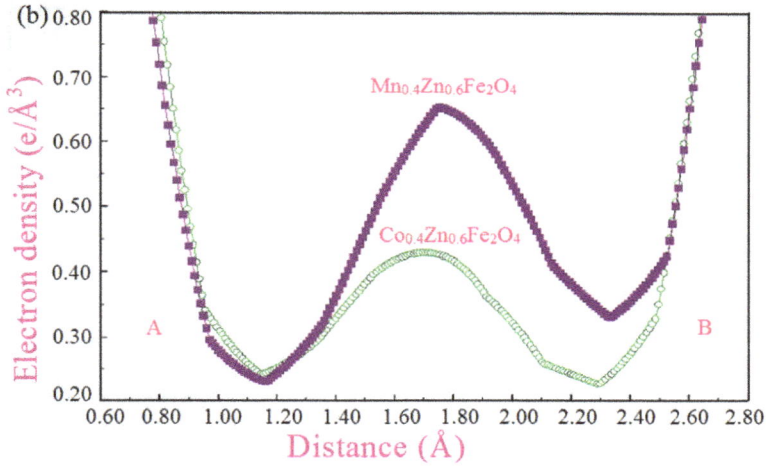

Fig.2.13(b) *One dimensional profile showing mid bond electron density values of tetrahedral (A) site - octahedral (B) site interactions of $Mn_{0.4}Zn_{0.6}Fe_2O_4$ and $Co_{0.4}Zn_{0.6}Fe_2O_4$ nano ferrites samples.*

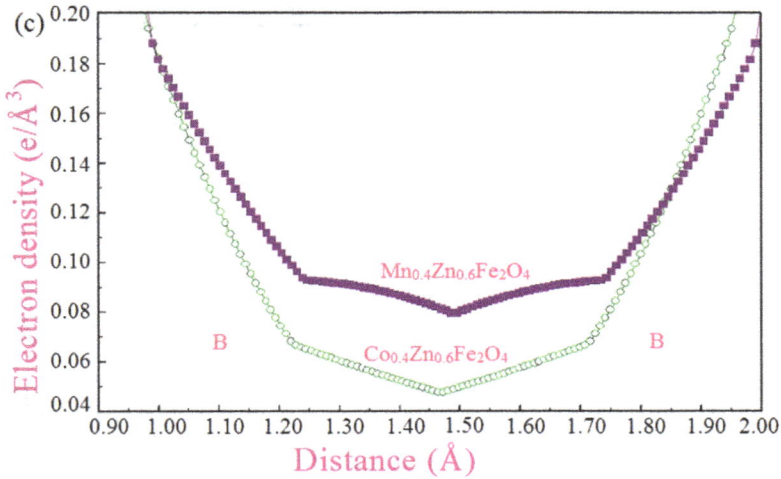

Fig.2.13(c) *One dimensional profile showing mid bond electron density values of octahedral (B) site - octahedral (B) site interactions of $Mn_{0.4}Zn_{0.6}Fe_2O_4$ and $Co_{0.4}Zn_{0.6}Fe_2O_4$ nano ferrites samples.*

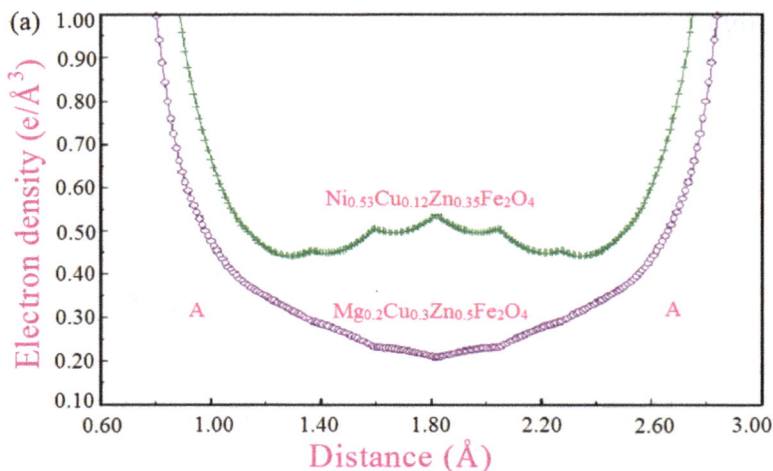

Fig.2.14(a) *One dimensional profile showing mid bond electron density values of tetrahedral (A) site - tetrahedral (A) site interactions of $Ni_{0.53}Cu_{0.12}Zn_{0.35}Fe_2O_4$ and $Mg_{.0.2}Cu_{0.3}Zn_{0.5}Fe_2O_4$ nano ferrites samples.*

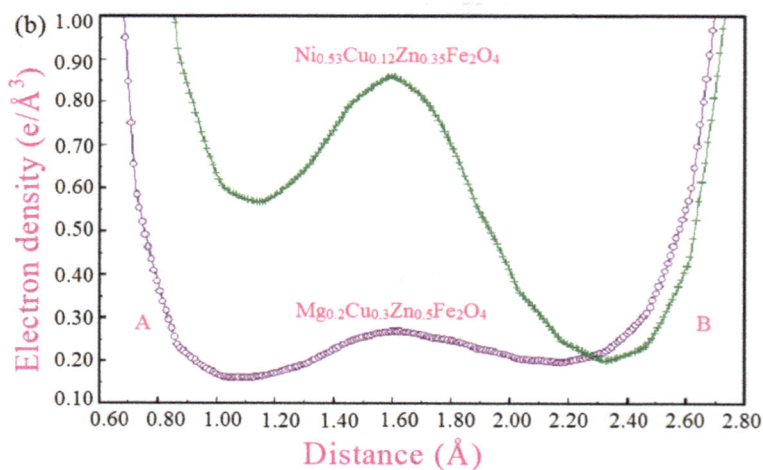

Fig.2.14(b) *One dimensional profile showing mid bond electron density values of tetrahedral (A) site - octahedral (B) site interactions of $Ni_{0.53}Cu_{0.12}Zn_{0.35}Fe_2O_4$ and $Mg_{.0.2}Cu_{0.3}Zn_{0.5}Fe_2O_4$ nano ferrites samples*

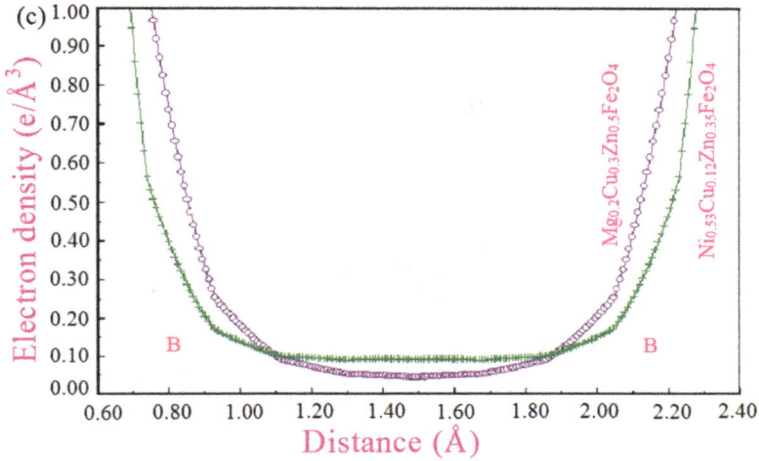

Fig.2.14(c) *One dimensional profile showing mid bond electron density values of octahedral (B) site - octahedral (B) site interactions of $Ni_{0.53}Cu_{0.12}Zn_{0.35}Fe_2O_4$ and $Mg_{0.2}Cu_{0.3}Zn_{0.5}Fe_2O_4$ nano ferrites samples.*

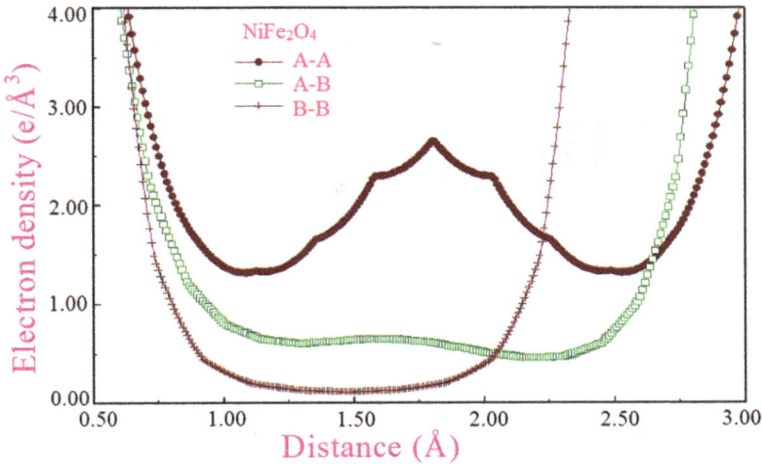

Fig.2.15 *One dimensional profile showing mid bond electron density values of tetrahedral (A) site - tetrahedral (A) site, tetrahedral (A) site - octahedral (B) site, octahedral (B) site - octahedral (B) site interactions of $NiFe_2O_4$ nano ferrites samples.*

Table 2.15 Mid bond electron density values of the samples obtained from MEM

Sample	Temp (°C)	A – A bond Mid bond electron density (e/Å³)	Distance (Å)	A – B bond Mid bond electron density (e/Å³)	Distance (Å)	B – B bond Mid bond electron density (e/Å³)	Distance (Å)
$Ni_{0.5}Zn_{0.5}Fe_2O_4$	700	0.668	1.82	0.521	1.70	0.118	1.48
	800	0.502	1.76	0.540	1.69	0.138	1.48
	900	0.456	1.78	0.584	1.70	0.153	1.48
	1000	0.393	1.80	0.495	1.67	0.129	1.48
	1100	0.390	1.80	0.454	1.72	0.113	1.48
	1200	0.496	1.81	0.508	1.74	0.078	1.48
	1300	0.383	1.81	0.611	1.73	0.074	1.48
	1400	0.446	1.81	0.566	1.73	0.103	1.48
$Ni_{0.8}Zn_{0.2}Fe_2O_4$	900	0.590	1.81	0.912	1.80	0.066	1.47
$Mn_{0.4}Zn_{0.6}Fe_2O_4$	900	0.426	1.82	0.653	1.75	0.079	1.49
$Co_{0.4}Zn_{0.6}Fe_2O_4$	900	0.191	1.79	0.428	1.71	0.049	1.46
$Ni_{0.53}Cu_{0.12}Zn_{0.35}Fe_2O_4$	900	0.525	1.79	0.857	1.60	0.092	1.48
$Mg_{0.2}Cu_{0.3}Zn_{0.5}Fe_2O_4$	900	0.207	1.82	0.267	1.60	0.044	1.48
$NiFe_2O_4$	700	2.644	1.80	0.643	1.63	0.119	1.47

2.8   Results - Magnetic properties

Ferrite being a magnetic material, the study of magnetic property gains importance in categorizing the materials such as dia, para, ferro magnetic material etc. The distribution of cations plays a crucial role in determining the various magnetic properties such as saturation magnetization (M_s), coercivity (H_c), strength of various site (A-A, A-B, B-B) interactions, presence of Yaffet-Kittel (YK) angle (Abdullah *et al.,* 2010) etc. in the unit cell of ferrite. Distribution of cations obtained from the result of the Rietveld refinement (Rietveld, 1969) can be established from the study of magnetic properties. The magnetic characterization results are analyzed for the effect of sintering temperature and substitution of various divalent cations on the various magnetic properties such as M_s, H_c, etc. The domain nature of the particles is also discussed. Figs.2.16(a-f) shows the relation between applied field and magnetization. The variation of saturation magnetization (M_S) with content of Fe ions at the A-site in the unit cell of $Ni_{0.5}Zn_{0.5}Fe_2O_4$ is show in Fig.2.17. Correlation between the coercivity and the grain size of $Ni_{0.5}Zn_{0.5}Fe_2O_4$ samples is shown in Fig.2.18. Various magnetic parameters estimated from hysteresis curve are tabulated in Table 2.16.

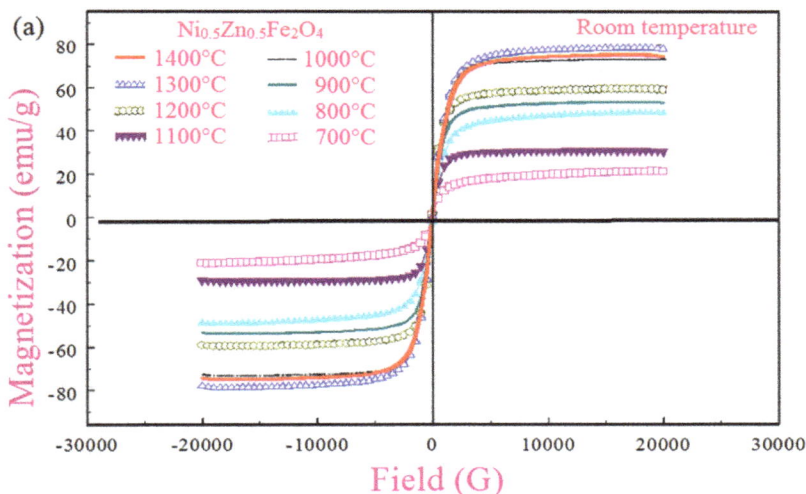

Fig.2.16(a) *Variation of magnetization with applied field for $Ni_{0.5}Zn_{0.5}Fe_2O_4$ samples sintered from 700 - 1400°C.*

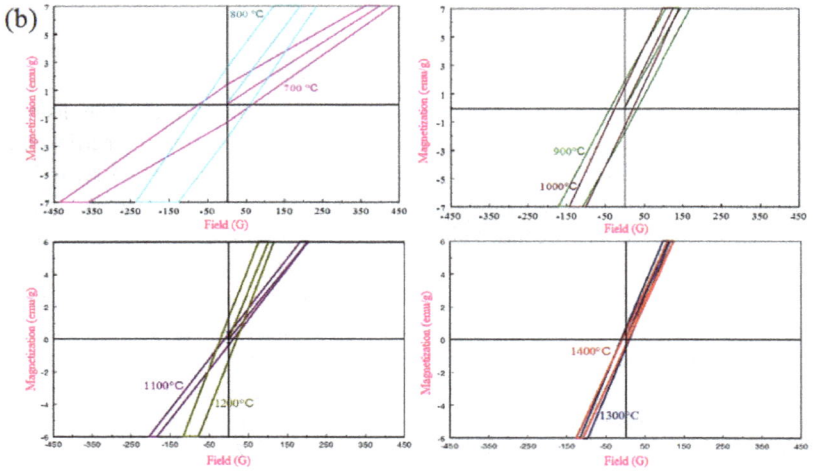

Fig.2.16(b) *Magnified view of variation of magnetization at lower applied field for $Ni_{0.5}Zn_{0.5}Fe_2O_4$ nano ferrite samples sintered from 700 - 1400°C.*

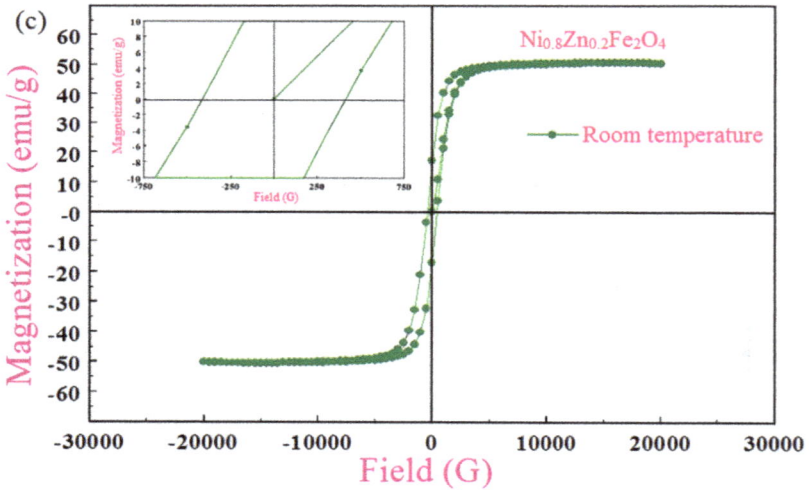

Fig.2.16(c) *Variation of magnetization with applied field for $Ni_{0.8}Zn_{0.2}Fe_2O_4$. Inset shows the magnified view at lower applied field.*

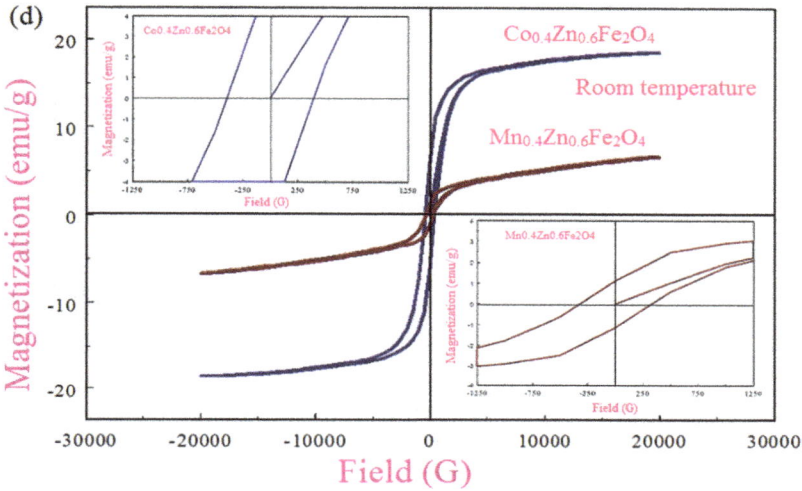

Fig.2.16(d) *Variation of magnetization with applied field for $Co_{0.4}Zn_{0.6}Fe_2O_4$ and $Mn_{0.4}Zn_{0.6}Fe_2O_4$. Inset shows the magnified view at lower applied field.*

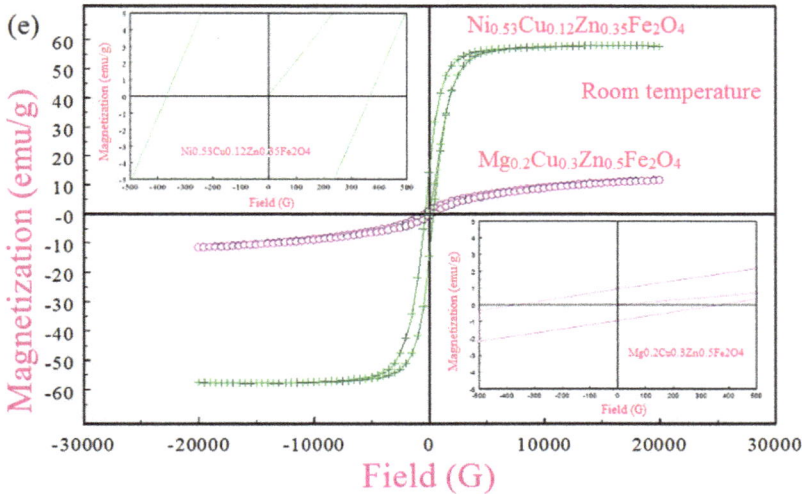

Fig.2.16(e) *Variation of magnetization with applied field for $Ni_{0.53}Cu_{0.12}Zn_{0.35}Fe_2O_4$ and $Mg_{0.2}Cu_{0.3}Zn_{0.5}Fe_2O_4$. Inset shows the magnified view at lower applied field.*

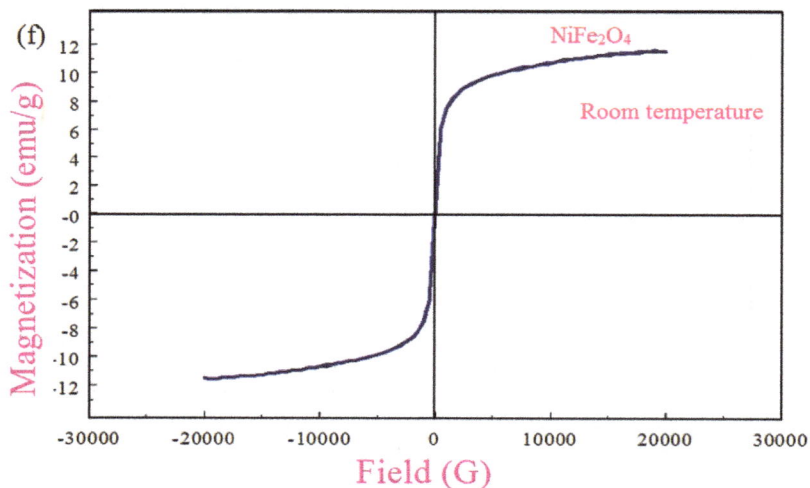

Fig.2.16(f) *Variation of magnetization with applied field for NiFe$_2$O$_4$.*

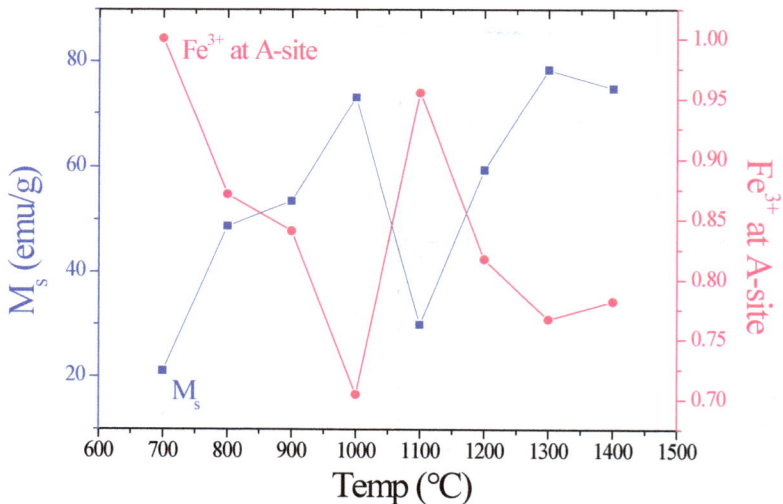

Fig.2.17 *Correlation between saturation magnetization (M_s) and Fe^{3+} content at A-site in the unit cell of ferrite.*

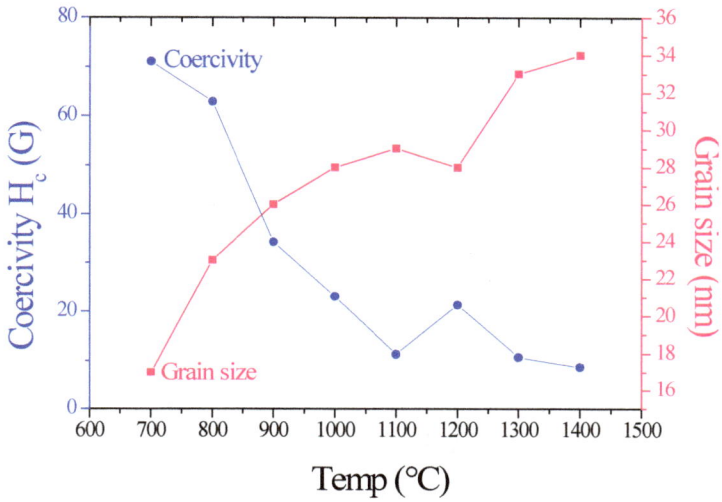

Fig.2.18 *Correlation between coercivity and the grain size of the $Ni_{0.5}Zn_{0.5}Fe_2O_4$ samples.*

Table 2.16 Magnetic parameters of synthesized samples.

Sample	Temp (°C)	M_s emu/g	Field at M_s ×10^3 (G)	H_{ci} G	M_r emu/g	Bohr Magneton μ_B^H	Bohr Magneton μ_B^N	Y-K angle (°)	$\mu \left(\frac{emu}{gkOe}\right)$	$\frac{M_r}{M_s}$
$Ni_{0.5}Zn_{0.5}Fe_2O_4$	700	21.11	19.5	71.07	1.37	0.89	1.00	10	1.53	0.065
	800	48.67	19.0	62.90	2.54	2.07	2.07	--	3.20	0.052
	900	53.52	18.5	34.22	1.78	2.28	2.27	--	4.13	0.033
	1000	73.14	17.0	23.09	1.36	3.11	3.12	--	4.67	0.019
	1100	29.98	15.5	11.35	0.36	1.27	1.27	--	2.53	0.012
	1200	59.44	17.5	21.32	1.35	2.83	2.55	16	5.04	0.023
	1300	78.43	16.0	10.72	0.61	3.38	3.38	--	4.53	0.008
	1400	74.88	17.0	8.73	0.45	3.18	3.18	--	4.10	0.006
$Ni_{0.8}Zn_{0.2}Fe_2O_4$	900	50.55	15.0	412.80	17.18	2.13	1.60	--	3.31	0.340
$Mn_{0.4}Zn_{0.6}Fe_2O_4$	900	6.66	19.5	324.29	7.33	0.28	2.00	41	0.27	0.170
$Co_{0.4}Zn_{0.6}Fe_2O_4$	900	18.62	19.5	395.25	0.93	0.80	3.81	53	1.21	0.320
$Ni_{0.53}Cu_{0.12}Zn_{0.35}Fe_2O_4$	900	58.03	16.0	367.50	14.40	2.48	5.62	52	3.12	0.250
$Mg_{0.2}Cu_{0.3}Zn_{0.5}Fe_2O_4$	900	11.58	20.0	364.23	0.96	0.48	6.91	75	0.21	0.080
$NiFe_2O_4$	700	11.60	19.0	3.90	0.05	0.48	2.39	42	0.98	0.008

2.9   Results - Optical properties

The band gap energy of the synthesized ferrite samples is determined, from the results of UV-visible characterization using Eq.1.60. A graph, as shown in Figs.2.19(a-n), is plotted between the energy of the incident photon (hv) as abscissa and square of the product of the absorption coefficient and the incident photon energy $(\alpha hv)^2$ as ordinate. Among the possible transitions, the allowed direct transition gives a straight line in all the synthesized samples. The extrapolation of straight line to the abscissa (y = 0) gives the value of band gap energy of the powder ferrite material. The band gap energy values of the samples synthesized in the present research study are tabulated in Table 2.17.

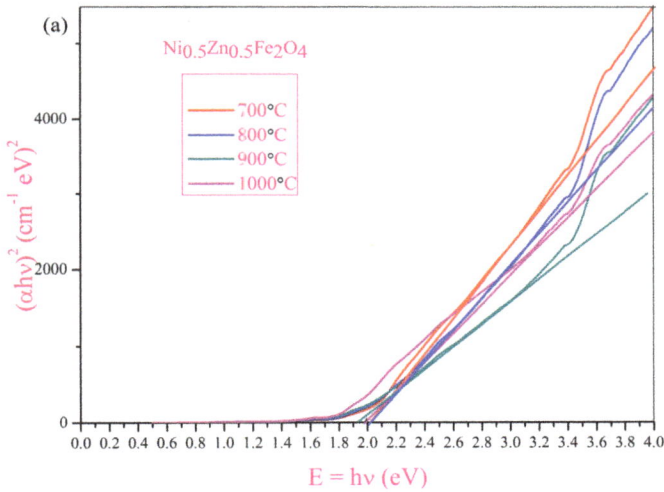

Fig.2.19(a) *Plot of energy (E) Vs $(\alpha E)^2$ of $Ni_{0.5}Zn_{0.5}Fe_2O_4$ nano ferrites samples sintered from 700 to 1000°C.*

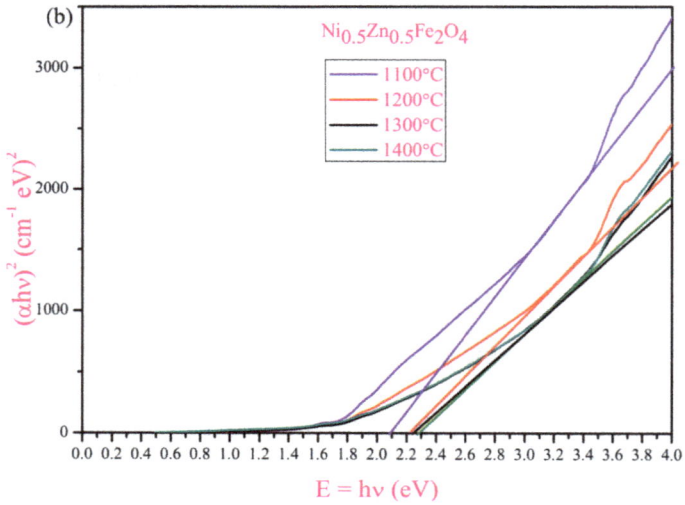

Fig.2.19(b) *Plot of energy (E) Vs $(\alpha E)^2$ of $Ni_{0.5}Zn_{0.5}Fe_2O_4$ nano ferrites samples sintered from 1100 to 1400°C.*

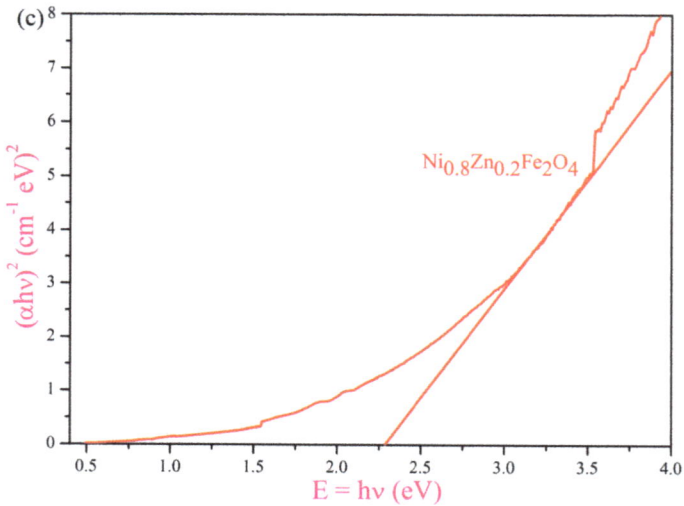

Fig.2.19(c) *Plot of energy (E) Vs $(\alpha E)^2$ of $Ni_{0.8}Zn_{0.2}Fe_2O_4$ nano ferrites.*

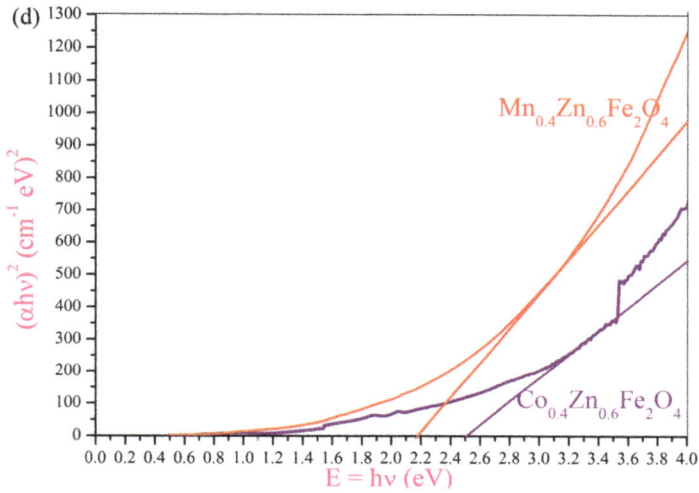

Fig.2.19(d) *Plot of energy (E) Vs (αE)^2 of Mn$_{0.4}$Zn$_{0.6}$Fe$_2$O$_4$ and Co$_{0.4}$Zn$_{0.6}$Fe$_2$O$_4$ nano ferrites samples.*

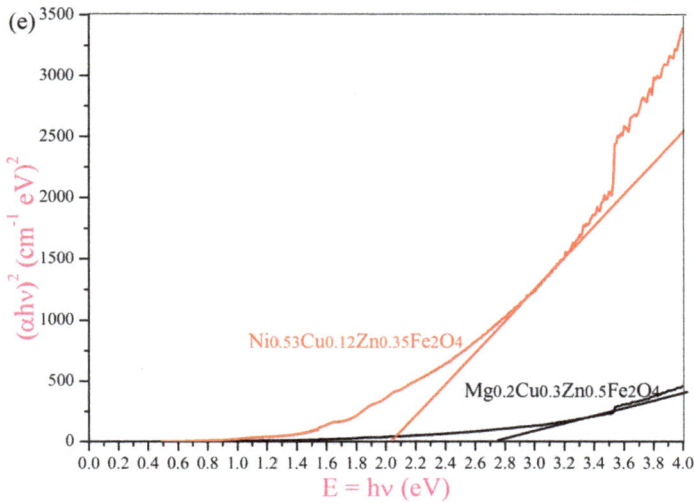

Fig.2.19(e) *Plot of energy (E) Vs (αE)^2 of Ni$_{0.53}$Cu$_{0.12}$Zn$_{0.35}$Fe$_2$O$_4$ and Mg$_{0.2}$Cu$_{0.3}$Zn$_{0.5}$Fe$_2$O$_4$ nano ferrites samples.*

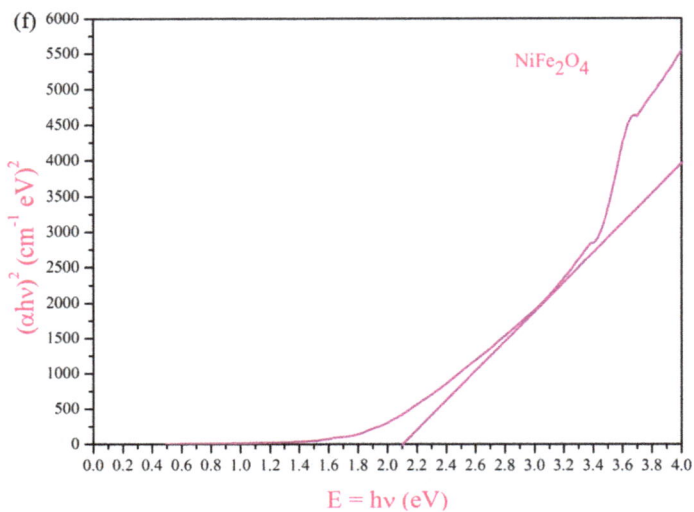

Fig.2.19(f) *Plot of energy (E) Vs (αE)² of NiFe₂O₄ nano ferrites.*

Table 2.17 *Optical band gap energy value of synthesized samples.*

Sample	Temp (°C)	Band gap energy value (eV)
$Ni_{0.5}Zn_{0.5}Fe_2O_4$	700	2.02
	800	2.00
	900	1.98
	1000	1.97
	1100	2.03
	1200	2.12
	1300	2.15
	1400	2.20
$Ni_{0.8}Zn_{0.2}Fe_2O_4$	900	2.29
$Mn_{0.4}Zn_{0.6}Fe_2O_4$	900	2.17
$Co_{0.4}Zn_{0.6}Fe_2O_4$	900	2.51
$Ni_{0.53}Cu_{0.12}Zn_{0.35}Fe_2O_4$	900	2.04
$Mg_{0.2}Cu_{0.3}Zn_{0.5}Fe_2O_4$	900	2.76
$NiFe_2O_4$	700	2.10

114

2.10 Results - Dielectric properties

The dielectric properties of ferrites depend on several factors including the chemical composition, method of preparation, sintering temperature and sintering atmosphere.

In the present research study, the dielectric properties such as real and imaginary dielectric constants, dielectric loss tangent and ac conductivity at room temperature in a frequency range of 0.1Hz to 15MHz on $Ni_{0.5}Zn_{0.5}Fe_2O_4$ samples sintered from 700 - 1200°C and $NiFe_2O_4$ are studied and the results are shown in Figs.2.20(a-d) and 2.21(a-b) and the corresponding various dielectric properties are tabulated in Tables 2.18 - 2.20 respectively.

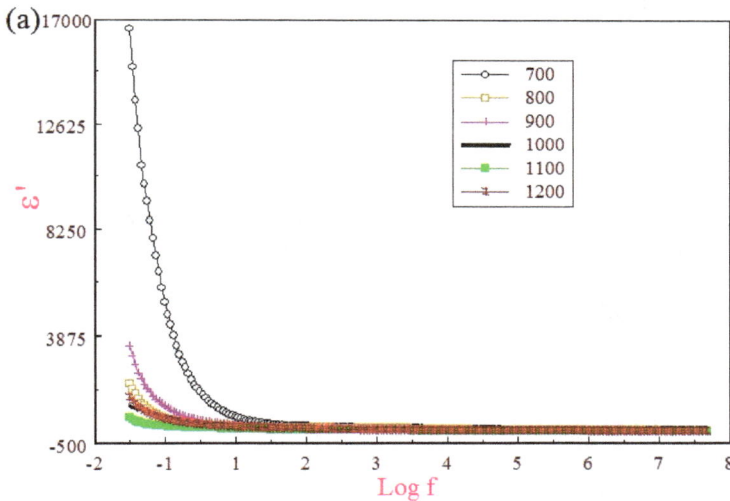

Fig.2.20(a) *Variation of real part of dielectric constant of $Ni_{0.5}Zn_{0.5}Fe_2O_4$ nano ferrite particles at room temperature with frequency in log scale.*

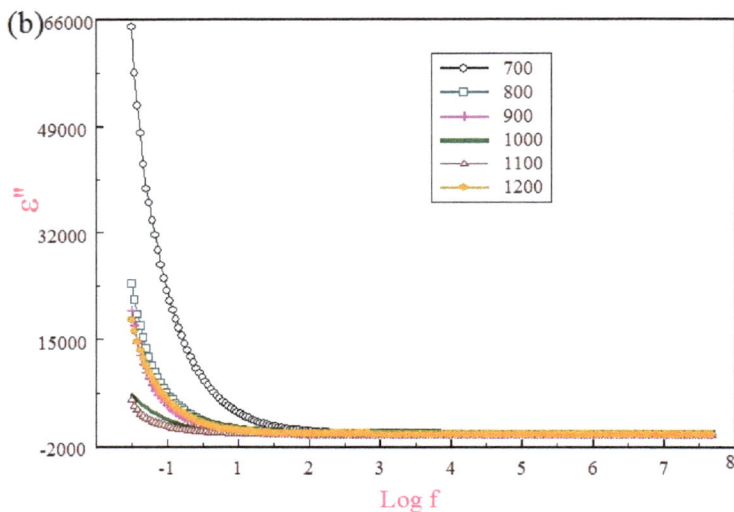

Fig.2.20(b) *Variation of imaginary part of dielectric constant of $Ni_{0.5}Zn_{0.5}Fe_2O_4$ nano ferrite particles at room temperature with frequency in log scale.*

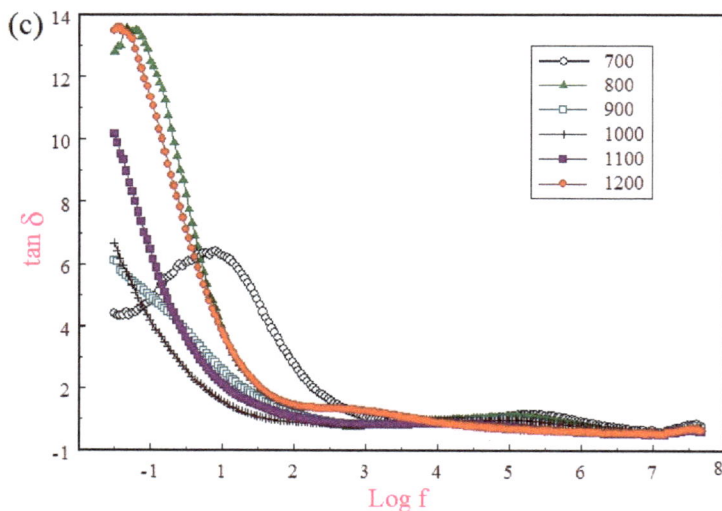

Fig.2.20(c) *Variation of dielectric loss factor of $Ni_{0.5}Zn_{0.5}Fe_2O_4$ nano ferrite particles at room temperature with frequency in log scale.*

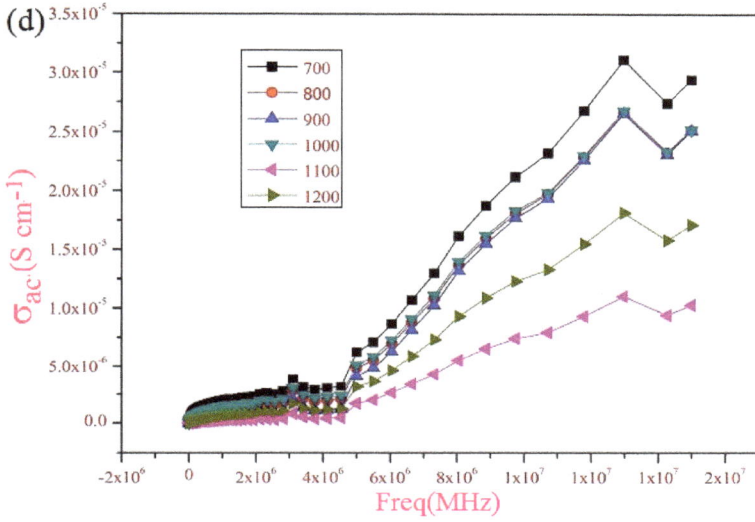

Fig.2.20(d) *Variation of AC conductivity of $Ni_{0.5}Zn_{0.5}Fe_2O_4$ nano ferrite particles at room temperature with frequency in log scale.*

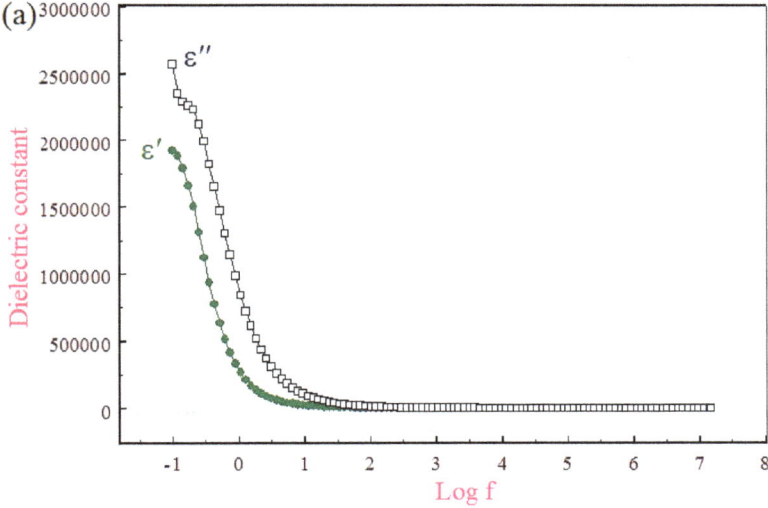

Fig.2.21(a) *Effect of frequency on ε' and ε'' at room temperature for $NiFe_2O_4$.*

117

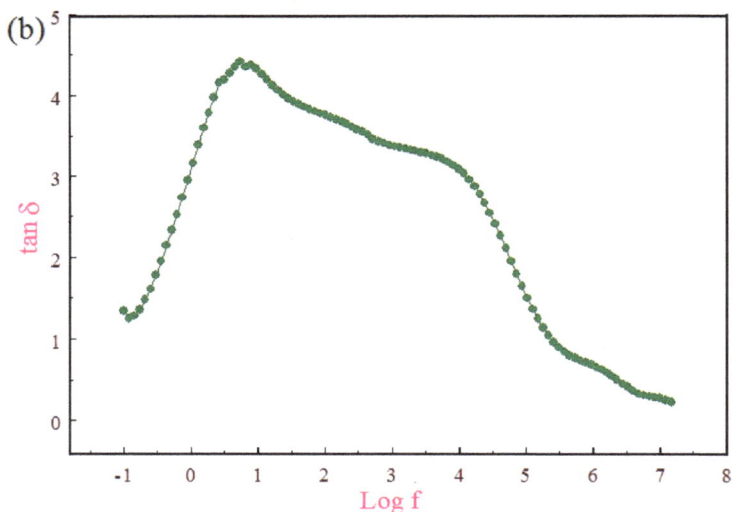

Fig.2.21(b) Plot of loss tangent against log frequency for NiFe$_2$O$_4$.

Table 2.18 *Various dielectric parameters of Ni$_{0.5}$Zn$_{0.5}$Fe$_2$O$_4$ ferrite particles.*

Sample Temp(°C)		At 0.1 Hz				At 15MHz			
		ε'	ε''	Tanδ	ρ (10^9)	ε'	ε''	Tan δ	ρ (10^4)
	700	16620	64842	3.90	0.28	11.8	3.52	0.30	3.40
	800	1933	23792	12.31	0.76	12.0	3.01	0.25	3.98
Ni$_{0.5}$Zn$_{0.5}$Fe$_2$O$_4$	900	3466	19499	5.63	0.92	11.9	3.00	0.25	3.99
	1000	1016	6269	6.17	2.87	14.5	3.01	0.21	3.97
	1100	564	5444	9.65	3.30	9.52	1.23	0.13	9.72
	1200	1389	18032	12.98	0.99	11.3	2.05	0.18	5.84

Table 2.19 *Thickness and radius value of Ni$_{0.5}$Zn$_{0.5}$Fe$_2$O$_4$ and NiFe$_2$O$_4$ ferrite particles.*

Sample	Temp(°C)	Thickeness (mm)	Diameter (mm)
	700	4.88	11.09
	800	4.39	10.99
Ni$_{0.5}$Zn$_{0.5}$Fe$_2$O$_4$	900	4.35	10.82
	1000	4.39	10.53
	1100	1.63	10.09
	1200	3.52	10.98
NiFe$_2$O$_4$	700	0.14	2.60

Table 2.20 *Various dielectric parameters of NiFe$_2$O$_4$ ferrite sample.*

Sample	At 0.1 Hz				At 15MHz			
	ε'	ε''	Tanδ	ρ	ε'	ε''	Tanδ	ρ
NiFe$_2$O$_4$	1.92×10^6	2.57×10^6	1.3	0.56×10^6	11	2.4	0.22	1.14×10^4

References

[1]   Abdullah DM, Khalid MB, Vivek V, Siddiqui WA, Kotnala RK (2010) J Alloys Compd 493:553-560. https://doi.org/10.1016/j.jallcom.2009.12.154

[2]   Beh HG, Lee KC, Hassan S (2014) Amer J Appl Sci 11(6): 878-882. https://doi.org/10.3844/ajassp.2014.878.882

[3]   Dalibor LS, Zorica ZL, Miljko VS, Cedomir DJ, Nebojsa ZR (2015) J Mater Sci:Mater Electron 26:1291-1303. https://doi.org/10.1007/s10854-014-2491-0

[4]   Deraz NM, Alarifi A (2012) Int J Electrochem Sci 7:3798 – 3808.

[5]   Gangatharan S, Chidambaram V, Kandasamy S (2010) Mate. Sci Appl 1:19-24.

[6]   Gull SF, Daniel GJ (1978) Nature 272:686-690. https://doi.org/10.1038/272686a0

[7]   Hankare PP, Sanadi KR, Garadkar KM, Patil DR, Mulla IS (2013) J Alloys Compd 553: 383–388. https://doi.org/10.1016/j.jallcom.2012.11.181

[8]   Izumi F, Dilanian RA (2002) Recent research developments in Physics part II Vol.3 Transworld Research Network, Trivandrum.

[9]   Momma K, Izumi F (2006) Comm Crystallogr Comput IUCr Newslett 7:106-119.

[10]   Petřiček V Dušek M, Palatinus L (2006) JANA2006: The Crystallographic Computing System, Institute of Physics, Academy of Sciences of the Czech Republic Praha 2000.

[11]   Rietveld HM (1969) J Appl Crystallogr 2:65-71. https://doi.org/10.1107/S0021889869006558

[12]   Saravanan R GRAIN software (private communication)

[13]   Satalkar M, Kane SN (2016) J Phys Conf Ser 755:012050 DOI:10.1088/1742-6596/755/1/012050. https://doi.org/10.1088/1742-6596/755/1/012050

[14]   Sidra Z, Shahid A, Saira R, Shahzad N (2016) Superlattices Microstruct 93:50-56. https://doi.org/10.1016/j.spmi.2016.02.048

Chapter 3

Discussion

Abstract

This chapter brings out a detailed discussion on the experimental lattice parameters, theoretical lattice parameters and oxygen positional parameters, radii of tetrahedral and octahedral sites evaluated from the results of powder XRD profile fitting and from the knowledge of cation distribution in the ferrite unit cell. Discussion on the mid bond electron density between tetrahedral and octahedral sites evaluated from maximum entropy method are presented and these results are compared with hitherto existing theoretical results. Magnetic properties, optical properties and dielectric properties of the synthesized ferrite samples are also discussed in this chapter.

Keywords

Structural Parameters, TG/DTA, Magnetic Properties, Optical Properties, Dielectric Properties, Cation Distribution

Contents

3.1   Discussion - Structural parameters of synthesized samples 122

3.1.1 Phase identification in the synthesized samples from
raw powder XRD 122

3.1.2 Cation distribution in the unit cell of synthesized samples 123

3.1.3 Discussion on refinement of raw XRD 123

3.1.3.1 Experimental lattice parameter of synthesized samples 124

3.1.3.2 Oxygen positional parameter of synthesized samples 125

3.1.3.3 Other structural parameters of synthesized samples 126

3.1.3.4 Theoretical lattice parameter of synthesized samples 128

3.1.3.5 Variation between theoretical and experimental lattice
parameter in $Ni_{0.5}Zn_{0.5}Fe_2O_4$ samples 128

3.2   Discussion on TG/DTA analysis of $Ni_{0.5}Zn_{0.5}Fe_2O_4$ (700 to 1000°C) and $Ni_{0.8}Zn_{0.2}Fe_2O_4$ sample..128

3.3   Discussion on morphological studies of synthesized samples............129

3.3.1   Discussion on SEM images..129

3.3.2   Discussion on EDAX results..130

3.3.3   Discussion on XRF results..131

3.4   Discussion on MEM results..131

3.4.1   Theories regarding various sites interactions in ferrite unit cell 131

3.4.2   Results based on theory regarding various site interactions in ferrite unit cell..132

3.4.3   MEM results on various sites interactions in ferrite unit cell..........133

3.4.3.1 $Ni_{0.5}Zn_{0.5}Fe_2O_4$ samples 134

3.4.3.2 $Ni_{0.8}Zn_{0.2}Fe_2O_4$ sample..135

3.4.3.3 $Mn_{0.4}Zn_{0.6}Fe_2O_4$ and $Co_{0.4}Zn_{0.6}Fe_2O_4$ samples................................135

3.4.3.4 $Ni_{0.53}Cu_{0.12}Zn_{0.35}Fe_2O_4$ and $Mg_{0.2}Cu_{0.3}Zn_{0.5}Fe_2O_4$ samples............136

3.4.3.5 $NiFe_2O_4$ sample 136

3.5   Discussion on magnetic properties...136

3.5.1   Saturation magnetization of the synthesized samples from the hysteresis curve..137

3.5.2   Coercivity of the synthesized samples from the hysteresis curve......138

3.5.3   Evaluation of magnetic moment from saturation magnetization and cation distribution data..140

3.6   Discussion on optical properties 142

3.7   Discussion on dielectric properties...143

3.7.1   Variation of dielectric constant with frequency of $Ni_{0.5}Zn_{0.5}Fe_2O_4$ samples..143

3.7.2   Variation of tanδ with frequency of $Ni_{0.5}Zn_{0.5}Fe_2O_4$ samples............144

3.7.3 Variation of AC conductivity with frequency of
Ni$_{0.5}$Zn$_{0.5}$Fe$_2$O$_4$ samples...145

3.7.4 Variation of dielectric constant with frequency of NiFe$_2$O$_4$ samples145

References 147

3.1 Discussion - Structural parameters of synthesized samples

The characteristics of a ferrite material depend on the physical properties such as the size and shape of the particles, composition and concentration of dopant ions, and the magnetic interactions between the A and B site ions in the unit cell. The interplay between the A and B site magnetic interactions often lead to novel and complex magnetic phenomena. In this chapter, various properties of the synthesized samples like structural, morphological, magnetic, optical and dielectric properties along with the results obtained from maximum entropy method are discussed.

3.1.1 Phase identification in the synthesized samples from raw powder XRD

The formation of spinel structure with Fd-3m (227) space group for all samples synthesized in this present research work, was confirmed with the presence of the following Bragg planes (202), (311), (222), (400), (422), (511) and (440) in the raw X-ray diffractogram, shown in Figs.2.1(a-e). Additional phases are also present in all but the NiFe$_2$O$_4$ samples.

The presence of a peak at 26.9° in samples sintered at 900 - 1400°C of Ni$_{0.5}$Zn$_{0.5}$Fe$_2$O$_4$ (Fig.2.1(a)) and Co$_{0.4}$Zn$_{0.6}$Fe$_2$O$_4$ (Fig.2.1(c)) confirm the presence of ε-Fe$_2$O$_3$ phase (JCPDS-160653) as an additional phase. The orthorhombic ε-Fe$_2$O$_3$ phase is an intermediate phase between α-Fe$_2$O$_3$ and γ-Fe$_2$O$_3$. Kelm and Mader (Kelm and Mader, 2006) have reported the formation of the ε-Fe$_2$O$_3$ phase at different preparation temperatures.

The presence of a peak at 80.9° in samples sintered at 700 - 1400°C of Ni$_{0.5}$Zn$_{0.5}$Fe$_2$O$_4$, Ni$_{0.8}$Zn$_{0.2}$Fe$_2$O$_4$, Mn$_{0.4}$Zn$_{0.6}$Fe$_2$O$_4$, Co$_{0.4}$Zn$_{0.6}$Fe$_2$O$_4$ (Figs.2.1(a-c)) and Mg$_{0.2}$Cu$_{0.3}$Zn$_{0.5}$Fe$_2$O$_4$ (Fig.2.1(d)) samples indicates the presence of a α-Fe$_2$O$_3$ phase (JCPDS-898103) as an additional phase. Various authors (Sharma et al., 2003; Sheikh et al., 2008; Barati, 2009; Gabal et al., 2010; Zaki et al., 2015) have also reported the presence of the α-Fe$_2$O$_3$ phase in ferrite samples.

3.1.2 Cation distribution in the unit cell of synthesized samples

The X-ray intensity ratio values of I_{220}/I_{400}, I_{220}/I_{440}, I_{422}/I_{440} and I_{400}/I_{440} planes and the cation distributions of all synthesized samples obtained in the present research work are tabulated in Table 2.1 and Table 2.2 respectively.

It can be seen that from Table 2.1, the close agreement between the calculated and observed intensity ratios asserts the cation distribution in the synthesized samples as shown in Table 2.2. In $Ni_{0.5}Zn_{0.5}Fe_2O_4$, $Ni_{0.8}Zn_{0.2}Fe_2O_4$ and $NiFe_2O_4$ samples, the ratio of observed and calculated intensities of I_{220}/I_{400}, I_{422}/I_{440} and I_{400}/I_{440} match very well. In $Mn_{0.4}Zn_{0.6}Fe_2O_4$ sample, the observed and calculated intensity ratio of I_{422}/I_{440} agrees very well, and in $Co_{0.4}Zn_{0.6}Fe_2O_4$ sample, the ratio of observed and calculated intensities of I_{220}/I_{400}, I_{422}/I_{440} and I_{400}/I_{440} planes agree very well. In the case of $Ni_{0.53}Cu_{0.12}Zn_{0.35}Fe_2O_4$ and $Mg_{0.2}Cu_{0.3}Zn_{0.5}Fe_2O_4$ samples the ratios of I_{220}/I_{440} and I_{422}/I_{440} ratios agree very well.

Table 2.2 indicates that cation disorder prevails in distribution of cations among the A and B site, i.e., cations occupying sites against their preferred sites such as occupation of zinc at B-site against its preferred A-site, occupation of nickel, magnesium and copper at A-site against their preferred B-site. Occupation of ions at their preferred site is strictly adhered in micrometer regime and not in nano meter regime. Disorder in cation distribution in ferrite nano particles is reported by various authors (Jeyadevan et al., 1994; Rath et al., 2000; Sreeja et al., 2008; Gabal et al., 2010; Singh et al., 2010; Slatineanu et al., 2011; Zaki et al, 2015).

The intensity of (220) and (440) planes are sensitive to the distribution of cations at A-site and B-site respectively in the unit cell. An increase in (220) plane intensity may be due to the migration of iron ions (Fe^{3+}) from B-site to A-site (Gangatharan et al., 2010) and the same kind of trend is observed in the samples $Ni_{0.5}Zn_{0.5}Fe_2O_4$ sintered from 700 - 1400°C, except the sample sintered at 900°C, as shown in Fig.2.2 and for the samples $Mn_{0.4}Zn_{0.6}Fe_2O_4$ and $Co_{0.4}Zn_{0.6}Fe_2O_4$ the correlation between I_{220} and Fe content at A-site is shown in Table.2.3.

3.1.3 Discussion on refinement of raw XRD

The raw XRD data of all the synthesized samples have been refined using the Rietveld refinement method by employing the software JANA 2006 software (Petříček et al., 2006) and the fitted profiles of the observed and calculated relative intensities, with the Bragg peaks together with their difference are shown in Figs.2.3(a-n). The reliability indices obtained from the Rietveld refinement method (Rietveld, 1969) using the JANA 2006 (Petříček et al., 2006) software for samples synthesized in the present research work are tabulated in Table 2.4. During refinement, various parameters like structural, lattice, peak shift, background profile shape and preferred orientation parameters were used to minimize the difference between a calculated and observed data. Minimal of F_o-F_c (Figs.2.3(a-n)) reflect the goodness of refinement for the synthesized samples besides the close agreement between the observed structure factor (F_o) and calculated structure factor (F_c) as tabulated in Tables 2.5(a – c).

The structural parameters, such as lattice parameters (a_{exp}), oxygen positional parameters (u), A-site and B-site radii ($r_{A(XRD)}$, $r_{B(XRD)}$), shared edge length of A and B-site (d_{AES}, d_{BES}) and unshared edge length of B-site (d_{BEU}) of the samples are tabulated in Table 2.6. The ferrite samples synthesized in the present research work belong to the spinel phase cubic crystal structure with space group Fd-3m. Hence, a_{exp} = b_{exp} = c_{exp} and the interfacial angles ($\alpha = \beta = \gamma$) are equal to 90°.

3.1.3.1   Experimental lattice parameter of synthesized samples

To analyze the experimental lattice parameters value of the synthesized samples, a comparision is made between (a_{exp}) and the ones reported in Joint Committee on Powder Diffraction Standards (JCPDS) of the corresponding samples and the same is tabulated in Table 2.7. The diffraction angle of 311 plane ($2\theta_{311}$) of the synthesized samples and the ones reported in JCPDS of the corresponding samples is also tabulated in Table 2.7.

It can be seen from Table 2.7 that the lattice parameter (a_{exp}) of the synthesized nano samples is larger than reported in the JCPDS, except for the samples $Co_{0.4}Zn_{0.6}Fe_2O_4$, $Mn_{0.4}Zn_{0.6}Fe_2O_4$ and $Mg_{0.2}Cu_{0.3}Zn_{0.5}Fe_2O_4$. Naughton et al., (Naughton et al., 2007) and Shahbaz et al., (Shahbaz et al., 2012) reported that the smaller the $2\theta_{311}$ values of the synthesized samples, the larger is its lattice parameter.

The variation in a_{exp} of the synthesized samples can be correlated to (i) the radius of substituting divalent ions and (ii) the distribution of cations among the A-site and B-site.

Since the composition remains the same in $Ni_{0.5}Zn_{0.5}Fe_2O_4$ series of samples, the variation in a_{exp} is attributed to the variation in A-site radius ($r_{A(XRD)}$). The value of $r_{A(XRD)}$ decreases with increase in sintering temperature up to 900°C and from 1000°C the value of $r_{A(XRD)}$ starts to increase except 1300°C. The variation in the value of a_{exp} follows the trend in the value of $r_{A(XRD)}$ except the sample sintered at 1100°C. The decrease in the a_{exp} value of the sample sintered at 1100°C is attributed to the secondary phase. The variation of a_{exp} with $r_{A(XRD)}$ is shown in Fig.2.4.

In the case of $Ni_{0.8}Zn_{0.2}Fe_2O_4$ sample, the larger a_{exp} value is attributed to the larger radius value of the substituting (r_{Zn}^{2+} = 0.74Å) zinc ions for nickel ions (r_{Ni}^{2+} = 0.69Å).

The diffraction angles of various Bragg planes of $Co_{0.4}Zn_{0.6}Fe_2O_4$ sample are larger than that of $Mn_{0.4}Zn_{0.6}Fe_2O_4$ and hence, the experimental lattice parameter a_{exp} of $Co_{0.4}Zn_{0.6}Fe_2O_4$ is smaller than that of $Mn_{0.4}Zn_{0.6}Fe_2O_4$.

Radius of manganese ion (r_{Mn} = 0.82Å) (Azaroff, 2009) is larger than zinc ion radius (r_{Zn}^{2+} = 0.74Å) (Azaroff, 2009) and the $2\theta_{311}$ value is less than that of bulk and hence, a larger a_{exp} value is expected in the case of $Mn_{0.4}Zn_{0.6}Fe_2O_4$ but the a_{exp} value is almost the same as the reported one in JCPDS -221012 (Table 2.7).

Radius of cobalt ion (r_{Co} = 0.72Å) (Azaroff, 2009) is almost the same as zinc ion radius (r_{Zn}^{2+} = 0.74Å) (Azaroff, 2009). The $2\theta_{311}$ value of synthesized $Co_{0.4}Zn_{0.6}Fe_2O_4$ sample is smaller than that of JCPDS -221012 (Table 2.7) and hence, a larger a_{exp} value is expected in $Co_{0.4}Zn_{0.6}Fe_2O_4$ sample but the observed a_{exp} value is lesser than the one reported in JCPDS -221012 (Table 2.7).

Thus, in $Mn_{0.4}Zn_{0.6}Fe_2O_4$ and in $Co_{0.4}Zn_{0.6}Fe_2O_4$ samples, the effect of $2\theta_{311}$ value and radius of the substituting ions are overshadowed by the presence of α-Fe_2O_3 phase and zinc loss during the synthesizing stage. Similar kind of observation is reported by Verma et al., (Verma et al., 2006) and Sheikh et al., (Sheikh et al., 2008) in Ni-Zn ferrite.

Similarly, the diffraction angles of various Bragg planes of $Mg_{0.2}Cu_{0.3}Zn_{0.5}Fe_2O_4$ are smaller than those of $Ni_{0.53}Cu_{0.12}Zn_{0.35}Fe_2O_4$ samples (Table 2.7), thus the lattice parameter (a_{exp}) of $Mg_{0.2}Cu_{0.3}Zn_{0.5}Fe_2O_4$ is larger than that of $Ni_{0.53}Cu_{0.12}Zn_{0.35}Fe_2O_4$. The $2\theta_{311}$ values of $Ni_{0.53}Cu_{0.12}Zn_{0.35}Fe_2O_4$ and $Mg_{0.2}Cu_{0.3}Zn_{0.5}Fe_2O_4$ samples are smaller than those reported in JCPDS – 100325 and JCPDS – 221012 respectively and hence, increase in the a_{exp} value is expected in both samples, but a smaller a_{exp} value in $Mg_{0.2}Cu_{0.3}Zn_{0.5}Fe_2O_4$ sample is observed. The decrease in a_{exp} value is explained on the basis of radius of the substituted ions. Substitution of Mg^{2+} (r_{Mg} = 0.66Å) (Azaroff, 2009) and Cu^{2+} (r_{Cu} = 0.72Å) (Azaroff, 2009) for Zn^{2+} (r_{Zn} = 0.74Å) (Azaroff, 2009) causes the a_{exp} value to decrease in $Mg_{0.2}Cu_{0.3}Zn_{0.5}Fe_2O_4$ sample. Substitution of Zn^{2+} (r_{Zn} = 0.74Å) (Azaroff, 2009) and Cu^{2+} (r_{Cu} = 0.72Å) (Azaroff, 2009) for Ni^{2+} (r_{Ni} = 0.69Å) (Azaroff, 2009) in $Ni_{0.53}Cu_{0.12}Zn_{0.35}Fe_2O_4$ sample causes the a_{exp} value to increase.

The increase in the a_{exp} value of $NiFe_2O_4$ sample is attributed to the shift of $2\theta_{311}$ value towards the lower value.

3.1.3.2   Oxygen positional parameter of synthesized samples

Oxygen positional parameter 'u' of all the synthesized samples is tabulated in Table 2.6. The ideal value of u in spinel structures is 0.375Å which indicates the cubic close packed arrangement of oxygen ions and in most spinel ferrites it is greater than 0.375 Å (Haralkar et al., 2013). The 'u' value obtained in the present research study is slightly larger, which indicates that the oxygen ions move away from the A-site cations along the [111] directions due to which the volume of each A-site interstices increases while the B-site decreases correspondingly (Chhaya *et al.*, 2011) and hence, the variations in the value of $r_{A(XRD)}$ and $r_{B(XRD)}$.

The variation in the d_{AES}, d_{BES}, d_{BEU} is attributed to the variation in $r_{A(XRD)}$ and $r_{B(XRD)}$. Whenever $r_{A(XRD)}$ decreases, it causes the radii of A-site ions to contract so that O^{2-} ions move close to A-site cations and hence, decrease in $r_{A(XRD)}$ and u. The variation in $r_{B(XRD)}$ is vice-versa to r_A(Satalkar *et al.*, 2016).

3.1.3.3 Other structural parameters of synthesized samples

Table 2.8 tabulates the unit cell volume (V), X-ray density (D_x) and the average grain size (t) of the synthesized samples.

The volume ($V = a_{exp}^3$) of the unit cell is estimated from the experimental lattice parameter a_{exp}. X-ray density (D_x) is evaluated using Eq.1.37, for all the synthesized samples. Eq.1.38 is used in evaluating average grain size 't' of the synthesized samples.

Since the molecular weight and composition remains the same for $Ni_{0.5}Zn_{0.5}Fe_2O_4$ samples, the variation in the D_x value is attributed to the variation in a_{exp} value. The value of X-ray density (D_x) increases with increase in sintering temperature which indicates the densification of samples with increase in sintering temperature, except for the 1200 and 1400°C samples,. The decrease in (D_x) value for the 1200 and 1400°C samples is attributed to the increase in intragranular porosity (Batoo *et al.*, 2012) since the average grain size (t) of 1200 and 1400°C samples is almost the same as that of samples sintered at 1100 and 1300°C (Table 2.8).

The X-ray density (D_x) of $Ni_{0.8}Zn_{0.2}Fe_2O_4$ sample (5.395g/cm^3) is larger when compared with 5.369 g/cm^3 (JCPDS 44-1485) of $NiFe_2O_4$ sample. Substitution of zinc (65.39 amu, $r_{Zn}^{2+} = 0.74Å$ (Azaroff, 2009)) for nickel (58.69 amu, $r_{Ni}^{2+} = 0.69Å$ (Azaroff, 2009)) causes the molecular weight and a_{exp} of the synthesized sample to increase and the increase in molecular weight overtakes the increase in lattice parameter and hence, D_x increases.

The X-ray density (D_x) value of $Mn_{0.4}Zn_{0.6}Fe_2O_4$ and $Co_{0.4}Zn_{0.6}Fe_2O_4$ sample is 5.236 and 5.525g/cm^3 respectively. The corresponding D_x values of $Mn_{0.4}Zn_{0.6}Fe_2O_4$ and $Co_{0.4}Zn_{0.6}Fe_2O_4$ are smaller and larger respectively when compared with 5.325g/cm^3 (JCPDS 22-1012) of the $ZnFe_2O_4$ sample. Substitution of manganese (54.94amu, $r_{Mn}^{2+} = 0.82Å$ (Azaroff, 2009)) and cobalt (58.93amu, $r_{Co}^{2+} = 0.72Å$ (Azaroff, 2009)) for zinc (65.39 amu, $r_{Zn}^{2+} = 0.74Å$ (Azaroff, 2009)) causes the molecular weight of both synthesized samples to decrease. Also volume of both samples is less when compared with 601.45Å^3 (JCPDS 22-1012) of the $ZnFe_2O_4$ sample. Decreased X-ray density (D_x) indicates density is improved due to the substitution of manganese for zinc in $Mn_{0.4}Zn_{0.6}Fe_2O_4$ sample and increased X-ray density (D_x) of $Co_{0.4}Zn_{0.6}Fe_2O_4$ sample indicates existence of pores in $Co_{0.4}Zn_{0.6}Fe_2O_4$ sample (Varalaxmi et al., 2013).

The X-ray density (D_x) value of $Ni_{0.53}Cu_{0.12}Zn_{0.35}Fe_2O_4$ and $Mg_{0.2}Cu_{0.3}Zn_{0.5}Fe_2O_4$ samples is smaller when compared with that of $NiFe_2O_4$ (JCPDS 100325) and $ZnFe_2O_4$ (JCPDS 221012) respectively. For $Ni_{0.53}Cu_{0.12}Zn_{0.35}Fe_2O_4$ sample, the substitution of Cu^{2+} (63.55 amu, $r_{Cu}^{2+} = 0.72\text{Å}$(Azaroff, 2009)) and Zn^{2+}(65.39 amu, $r_{Zn}^{2+} = 0.74\text{Å}$ (Azaroff, 2009)) for Ni^{2+} (58.69 amu, $r_{Ni}^{2+} = 0.69\text{Å}$ (Azaroff, 2009)) increases the molecular weight of the sample, though this increase is not sufficient enough to offset the increase in volume of the sample and hence, X-ray density (D_x) decreases. In the case of $Mg_{0.2}Cu_{0.3}Zn_{0.5}Fe_2O_4$ sample, Zn^{2+} (65.39 amu, $r_{Zn}^{2+} = 0.74\text{Å}$ (Azaroff, 2009)) is substituted by lesser weight Cu^{2+} (63.55 amu, $r_{Cu}^{2+} = 0.72\text{Å}$ (Azaroff, 2009)) and Mg^{2+} (24.31 amu, $r_{Mg}^{2+} = 0.66\text{Å}$ (Azaroff, 2009)) hence, decrease in molecular weight and this decrease is sufficient enough to compensate the decrease in volume of the sample hence, X-ray density (D_x) decreases for $Mg_{0.2}Cu_{0.3}Zn_{0.5}Fe_2O_4$ sample (Varalaxmi et al., 2013).

The X-ray density (D_x) value of $NiFe_2O_4$ samples is smaller when compared with 5.37 g/cm^3 (JCPDS-100325).

For the $Ni_{0.5}Zn_{0.5}Fe_2O_4$ samples, the diffraction angle (in radians) and full width at half maximum (in radians) of the most intense Bragg planes, namely, (202), (311), (400), (422), (511) and (440) were used in calculating average crystallite size 't'. The increase in grain size is attributed to coalescence of particles that generally occurs with increase in sintering temperature and this increase in grain size leads to decrease in porosity (Abdullah et al., 2010) in the samples. Sharp increase in the grain size is observed for the samples sintered at 800 and 1300°C when compared with their predecessor, whereas for the samples sintered from 900-1200°C, the grain size almost remains uniform. Marginal increase in grain size of the sample sintered at 1100°C is attributed to the presence of a greater amount of second phase when compared to 1000°C.

For the sample $Ni_{0.8}Zn_{0.2}Fe_2O_4$, three most intense Bragg planes (311), (400) and (440) were used to determine the average grain size.

Bragg planes (311), (511) and (440) were employed for the $Ni_{0.53}Cu_{0.12}Zn_{0.35}Fe_2O_4$ and the $Mg_{0.2}Cu_{0.3}Zn_{0.5}Fe_2O_4$ samples to determine the grain sizes. The grain sizes of the samples indicate that the porosity of $Mg_{0.2}Cu_{0.3}Zn_{0.5}Fe_2O_4$ is less than that of $Ni_{0.53}Cu_{0.12}Zn_{0.35}Fe_2O_4$.

Among the synthesized samples, $NiFe_2O_4$ records the smallest 't' value (Table 2.6).

3.1.3.4 Theoretical lattice parameter of synthesized samples

Table 2.9 tabulates the a_{the}, $r_{A(CD)}$ and $r_{B(CD)}$ of the samples synthesized in the present research work. Comparing a_{the} with a_{exp} (Table 2.6) it can be seen that a_{exp} is larger than a_{the}, except for the sample $Co_{0.4}Zn_{0.6}Fe_2O_4$. The difference in a_{the} and a_{exp} attributed to the presence of Fe^{2+} ions at B-sites. At higher sintering temperature, zinc loss may occur and to maintain the charge neutrality, Fe^{3+}(r_{Fe}^{3+} = 0.64 Å (Azaroff, 2009)) ion reduced to Fe^{2+} (r_{Fe}^{2+} = 0.78 Å(Azaroff, 2009)). Amount of Fe^{2+} ions present were not taken into consideration for theoretical calculations and hence, the difference between a_{the} with a_{exp} (Varaprasad et al., 2015). Decrease in a_{exp} of $Co_{0.4}Zn_{0.6}Fe_2O_4$ sample is attributed to the combined effect of presence of additional phases and zinc loss during the synthesizing stage (Verma et al., 2006; Sheikh et al., 2008). The difference between $r_{A(XRD)}$ and $r_{A(CD)}$, $r_{B(XRD)}$ and $r_{B(CD)}$ can be explained on the same line of reasoning that attributed to the difference between a_{exp} and a_{the} (Kurmude et al., 2014).

3.1.3.5 Variation between theoretical and experimental lattice parameter in $Ni_{0.5}Zn_{0.5}Fe_2O_4$ samples

Fig.2.5a shows the relation between the experimental (a_{exp}) and theoretical (a_{the}) lattice parameter of $Ni_{0.5}Zn_{0.5}Fe_2O_4$ samples sintered from 700-1400°C. The trend in a_{exp} and a_{the} is not linear and the trend is akin only for the samples sintered between 900 - 1200°C. This observed variation between a_{exp} and a_{the} can be explained as follows. Eq.1.7 is applicable to specimens that have orientation in all possible orientations. However, all synthesized polycrystalline specimens have preferential orientation of the grains in a higher or lower degree (Varaprasad et al., 2015). The difference between a_{exp} and a_{the} decreases with increase in sintering temperature as shown in Fig.2.5b except for the samples 1200 and 1400°C.

3.2 Discussion on TG/DTA analysis of $Ni_{0.5}Zn_{0.5}Fe_2O_4$ (700 to 1000°C) and $Ni_{0.8}Zn_{0.2}Fe_2O_4$ sample

Figs. 2.6(a-e) show the TG-DTA analysis carried out for the samples, $Ni_{0.5}Zn_{0.5}Fe_2O_4$ sintered from 700 to 1000°C and $Ni_{0.8}Zn_{0.2}Fe_2O_4$ sample sintered at 900°C, in static air atmosphere in the temperature range of 40–730°C at the rate of 5°C/min.

In $Ni_{0.5}Zn_{0.5}Fe_2O_4$ samples, TG curve shows that there is an initial weight loss which is attributed to the evaporation of adsorbed water in the sample, followed by a peak whose width increases with increase in sintering temperature except for the sample sintered at 1000°C. These peaks around 100–200°C are attributed to the oxidation reaction at the surface of the sintered samples. All the samples thereafter recorded continuous weight loss which are ascribed to the decomposition of organic matrix and the same fact is established by the DTA curve by means of exothermic peaks around 300°C in all samples and the weight loss is almost stable thereafter indicating the onset of the crystallization process. The sample sintered at 700°C shows two exothermic peaks respectively at 291.13°C and 359.69°C and also an endothermic peak at 460.18°C and the sample sintered at 800°C shows an exothermic peak at 346.93°C and an endothermic peak at 467.88°C. Neither exothermic nor endothermic peaks are observed in the samples sintered at 900 and 1000°C. The total percentage of weight loss% in the sample sintered at 800°C and 900°C is 1% while the sample sintered at 700°C is 2% whereas the sample sintered at 1000°C recorded 2.5% weight loss.

The TG curve for $Ni_{0.8}Zn_{0.2}Fe_2O_4$ sample shows initial weight loss which is due to the evaporation of physically and chemically adsorbed water and the weight loss continues up to the maximum studied temperature. From the TG graph, it can be seen that the total weight loss in the studied temperature range (from 40 - 730°C) is only 2%. From the thermal behaviour, it can be ascertained that no anomalies are observed and the behaviour is as expected from already sintered samples.

3.3 Discussion on morphological studies of synthesized samples

3.3.1 Discussion on SEM images

The SEM pictures of the samples synthesized in this research work are shown in Figs.2.7(a–n).

The effect of sintering on the particle size of $Ni_{0.5}Zn_{0.5}Fe_2O_4$ samples sintered from 700-1400°C can be seen in the Figs.2.7(a-h). The presence of porosity and non-homogeneous distribution of particle sizes are observed from the scanning electron microscope images of the samples. A remarkable growth in particle size is observed with increase in sintering temperature.

Uniformly distributed particles can be seen from the morphology of the $Ni_{0.8}Zn_{0.2}Fe_2O_4$ sample as shown in Fig. 2.7(i).

SEM images of $Mn_{0.4}Zn_{0.6}Fe_2O_4$ and $Co_{0.4}Zn_{0.6}Fe_2O_4$ samples are shown in Figs.2.7(j-k) respectively. In the case of $Mn_{0.4}Zn_{0.6}Fe_2O_4$ sample, bigger size and rectangle like particles are clearly seen in Fig.2.7(j). Homogeneous distribution and fine crystalline nature of $Co_{0.4}Zn_{0.6}Fe_2O_4$ sample can be seen from Fig.2.7k. The broad and high intensity peaks in the raw XRD (Fig.2.1(c)) of $Co_{0.4}Zn_{0.6}Fe_2O_4$ sample manifest the same.

The broad and high intensity peaks in the raw XRD (Fig.2.1(d)) of $Ni_{0.53}Cu_{0.12}Zn_{0.35}Fe_2O_4$ sample than that of $Mg_{0.2}Cu_{0.3}Zn_{0.5}Fe_2O_4$ reveal the fine crystalline nature of $Ni_{0.53}Cu_{0.12}Zn_{0.35}Fe_2O_4$ sample, and the same is confirmed from the scanning electron microscope image of the samples which are shown in Figs.2.7(l-m). Homogeneous distribution of particles with well-defined grain boundary can be seen in the case of $Mg_{0.2}Cu_{0.3}Zn_{0.5}Fe_2O_4$ sample whereas in $Ni_{0.53}Cu_{0.12}Zn_{0.35}Fe_2O_4$, agglomeration of particles and presence of porosity can be seen.

Fig.2.7(n) shows the scanning electron microscope image of the $NiFe_2O_4$ sample. Random distribution in the particle sizes and presence of pores can be clearly seen.

Table 2.10 lists the comparison between the particle size estimated from SEM pictures of the synthesized samples and the estimation of grain size from XRD using the software GRAIN (Saravanan R, GRAIN software). In the GRAIN software, the grain size of the synthesized samples is estimated by using the full width half maximum of the powder XRD peaks of the corresponding samples. The number (N) of the coherently diffracting domains can be obtained by calculating the ratio of particle size from SEM to grain size from XRD (Saravanan *et al.,* 2010). For the sample $Ni_{0.5}Zn_{0.5}Fe_2O_4$ sample sintered at 700°C, N value is 23 which mean that there are approximately twenty three coherently diffracting domains in each particle (Guiner A, 1994). The value of N for the $Ni_{0.5}Zn_{0.5}Fe_2O_4$ samples sintered at 1300 and 1400°C is equal to 2500 and 2000 respectively. Since the powder XRD gives only the size of the coherently diffracting domains, it cannot be directly compared to the size obtained using an SEM except in rare cases.

3.3.2 Discussion on EDAX results

Figs.2.8(a-g) show the results of EDAX analysis. Energy dispersive X-ray spectrometry (EDAX) measurements were carried out on $Ni_{0.5}Zn_{0.5}Fe_2O_4$ samples sintered at 900, 1000, 1100 and 1200°C, $Ni_{0.8}Zn_{0.2}Fe_2O_4$, $Mn_{0.4}Zn_{0.6}Fe_2O_4$, and $NiFe_2O_4$ samples to analyze the estimated stoichiometry. The expected and estimated stiochiometry of the various elements that are present in $Ni_{0.5}Zn_{0.5}Fe_2O_4$ samples sintered at 900, 1000, 1100 and 1200°C, $Ni_{0.8}Zn_{0.2}Fe_2O_4$, $Mn_{0.4}Zn_{0.6}Fe_2O_4$ and $NiFe_2O_4$ samples are tabulated in Tables 2.11(a-d).

The expected stoichiometries of Ni/Fe, Zn/Fe and Ni/Zn in 900 and 1000°C samples are 0.25, 0.25 and 1 respectively. From the EDAX result, the estimated stoichiometric ratios correspondingly is found out as 0.28, 0.28 and 0.99 for 900°C and for 1000°C it is 0.27, 0.24 and 1.1. Similarly for 1100°C and 1200°C samples the expected ratios of Ni:Fe, Zn:Fe, and Ni:Fe for both samples are 0.25, 0.25, and 1 respectively, whereas the estimated ratio of Ni:Fe, Zn:Fe, and Ni:Fe comes out as 0.259, 0.260, and 0.995 for 1100°C and for 1200°C it is 0.256, 0.251 and 1.021 respectively. The estimated and expected stoichiometric ratios match closely and hence it is ascertained that the samples contain the elements in the expected stoichiometry. A peak corresponding to Yttrium (sample holder) is also observed.

EDAX analysis of the $Ni_{0.8}Zn_{0.2}Fe_2O_4$ shows the sample is not in the stiochiometric ratio. The expected stoichiometry ratio of Zn/Fe, Ni/Fe and Ni/Zn is 0.1, 0.4 and 4.0 respectively whereas the estimated ratio comes out as 0.15, 0.97 and 6.38 respectively and non-stiochiometric ratio is attributed to the loss of zinc ions during the preparation stage of the sample and to compensate the zinc loss, Fe^{3+} reduces to Fe^{2+}. Fe^{2+} ions prefer to occupy octahedral site forcing the Fe^{3+} ions to occupy the tetrahedral site which in turn may force the zinc to occupy octahedral sites. This may cause the radius of ions at A-site ($r_{A(XRD)}$) to decrease and that of B-site ($r_{B(XRD)}$) to increase, which is confirmed by the ($r_{A(XRD)}$) and ($r_{B(XRD)}$) value evaluated from XRD data (Table 2.6) and this is further confirmed by cation distribution (Table 2.2).

For the sample $Mn_{0.4}Zn_{0.6}Fe_2O_4$, the expected and estimated ratios of Mn/Fe, Zn/Fe and Mn/Zn agree well indicating that the sample is in stiochiometric ratio.

For $NiFe_2O_4$ sample, the expected ratio for nickel to iron is 0.5 whereas the estimated stoichiometry from EDAX analysis is 0.4.

3.3.3 Discussion on XRF results

The elemental composition of the samples is analyzed by X-ray fluorescence (XRF) method and the percentage concentrations of elements in the synthesized samples are tabulated in Tables 2.12(a-e). The purity of the $Ni_{0.5}Zn_{0.5}Fe_2O_4$ (1400°C), $Ni_{0.53}Cu_{0.12}Zn_{0.35}Fe_2O_4$ and $NiFe_2O_4$ samples are confirmed by detected elements listed in Table 2.12(a), 2.12(d) and 2.12(e) respectively. In $Ni_{0.8}Zn_{0.2}Fe_2O_4$ sample, traces of Mg and in $Mn_{0.4}Zn_{0.6}Fe_2O_4$ and $Co_{0.4}Zn_{0.6}Fe_2O_4$ samples, traces of Cu and Ni and in $Mg_{0.2}Cu_{0.3}Zn_{0.5}Fe_2O_4$ sample traces of Ni are found as listed in Table 2.12(b), 2.12(c) and 2.12(d) respectively.

3.4 Discussion on MEM results

3.4.1 Theories regarding various sites interactions in ferrite unit cell

The theoretical views, regarding the bond strength between the various sites interactions in the unit cell of ferrite are summarized in this section.

Lakhani et al. (Lakhani et al., 2011) have reported decrease in interionic distances between cation and anion ($M_e - O$) and between cations ($M_e - M_e$) and increase in the bond angles θ_1, θ_2 and θ_5 values and decrease in θ_3 and θ_4 values strengthens A-B and A-A interactions and weakens B-B interactions in aluminium doped in copper ferrite.

Gagankumar et al. (Gagankumar et al., 2015) have reported that the weakening of A-B interactions may be owed to the increase in interionic distances and interionic bond angles (M_e – O, M_e – M_e and θ_1, θ_2 and θ_5) and decrease in θ_3 and θ_4 values in gadolinium substituted magnesium-manganese ferrite.

Battu et al. (Battu et al., 2007) have reported that increase in interionic distances and increase in θ_1 and θ_2 bond angle values weakens A-B interactions and decrease in θ_3, θ_4 and θ_5 values strengthens B-B and A-A interactions in nickel substituted lithium ferrite.

3.4.2 Results based on theory regarding various site interactions in ferrite unit cell

Analyzing the results in Table 2.13(a), it can be seen that for the sample $Ni_{0.5}Zn_{0.5}Fe_2O_4$, the interionic distances (M_e – M_e) decrease with increase in sintering temperature up to 1100°C, thereafter the value shows an increasing trend except the sample sintered at 1300°C. In the case of sample sintered at 1000°C, the cation – cation (M_e – M_e) values are almost equal to that of sample sintered at 900°C.

Similarly, interionic distance (M_e – O) decreases with increase in sintering temperature up to 900°C, thereafter interionic distance shows increasing tendency except for the sample sintered at 1300°C. The opposite tendency in 'p', which is the distance between the B-site and the oxygen atom, is attributed to the drift observed in 'u'. An increase in 'u' indicates that the distance between A-site and oxygen atom increases and hence, the distance between the B-site and the oxygen atom 'p' decreases and vice-versa.

In the case of interionic bond angles, increase in θ_1, θ_2 and θ_5 and decrease in θ_3 and θ_4 are observed with increase in sintering temperature up to 900°C and the trend is reversed for the sample sintered at 1000°C. For the samples sintered at 1100 to 1400°C, an alternate decrease and increase is observed in θ_1, θ_2 and θ_5, whereas in θ_3 and θ_4 an alternate increase and decrease is observed.

Analyzing the (M_e-M_e), (M_e-O) and (θ_1-θ_5) values in the Table 2.13(a), for the sample sintered at 800, 900 and 1300°C strengthening of A-B, A-A and weakening of B-B interactions would occur with respect to that of their predecessor sample in accordance with the analysis of Lakhani et al. (Lakhani et al., 2011).

For the samples sintered at 1000, 1100, 1200 and 1400°C, weakening of A-B, A-A and strengthening of B-B interactions should occur with respect to that of their predecessor sample in accordance with the analysis of Lakhani et al. (Lakhani et al., 2011).

The sample $Co_{0.4}Zn_{0.6}Fe_2O_4$ (Table 2.13(b)) should have stronger A-B, A-A interaction and weaker B-B interaction than $Mn_{0.4}Zn_{0.6}Fe_2O_4$ sample in consistent with the result of Lakhani et al. (Lakhani et al., 2011).

Similarly, the sample $Ni_{0.53}Cu_{0.12}Zn_{0.35}Fe_2O_4$ (Table 2.13(c)), should have stronger A-B, A-A interaction and weaker B-B interaction than $Mg_{0.2}Cu_{0.3}Zn_{0.5}Fe_2O_4$ keeping in line with the analysis of Lakhani et al. (Lakhani et al., 2011).

3.4.3 MEM results on various sites interactions in ferrite unit cell

To validate various interactions on the synthesized ferrite samples, the maximum entropy method (MEM) (Gull and Daniel, 1978) was employed to evaluate the mid bond electron density of various interactions namely, A-A, A-B and B-B sites in the unit cell of ferrite. The PRIMA (Practice Iterative MEM Analyses) (Izumi and Dilanian, 2002) software, a MEM refinement program, was employed to calculate the mid bond electron densities from the X-ray diffraction data and the VESTA (Visualization of Electron/Nuclear densities and Structures) (Momma and Izumi, 2006) software was used to plot the 3D, 2D and 1D plot of electron densities in this research work. Table 2.14 tabulates the result of MEM refinement on the synthesized samples.

The electron density distribution in 3D isosurface for all samples synthesized in the present research work, is shown in Figs.2.9(a-n). 2D electron density contour maps on the Miller plane (110) of the samples are shown in Figs.2.10(a-n).

In ferrites, the spin interaction varies directly with bond angle and inversely with interionic distances (Lakhani *et al.,* 2011) and the A-B interaction takes place through super exchange interaction i.e., the atoms at A-sites and at B-sites interact through oxygen atoms. The interactions between A-site atoms and oxygen atom and similarly interactions between oxygen atom and B-site atoms can be clearly viewed in the Figs.2.10(a-n), through the presence of residual electron contours present around the A-site, B-site and oxygens atoms in (110) plane. In addition (i) A-A sites interactions are observed by the presence of dense residual electron clouds between two nearest A-A sites atoms in (110) plane and (ii) presence of thin residual electron cloud between two nearest B-B sites atoms in (110) plane when compared to A-A site interactions.

To obtain the numerical values for mid bond electron density of the A-B, A-A and B-B interactions of the respective samples, 1D electron density profiles are drawn as depicted in Figs.2.11(a-c), 2.12, 2.13(a-c), 2.14(a-c) and in Fig.2.15 and their values are tabulated in Table 2.15.

3.4.3.1 $Ni_{0.5}Zn_{0.5}Fe_2O_4$ samples

MEM results are in agreement with the theoretical assumption (Lakhani *et al.*, 2011) with respect to the A-B site interactions are concerned. From the mid bond electron density values of $Ni_{0.5}Zn_{0.5}Fe_2O_4$ samples, as tabulated in Table 2.15, it can be seen that strengthening of A-B interactions in the samples sintered at 800, 900 and 1300°C and weakening of A-B interactions in the sample sintered at 1000, 1100 and 1400°C samples have occurred with respect to that of their predecessor sample. MEM result of the sample sintered at 1200°C contradicts with theoretical assumption (Lakhani *et al.*, 2011) by recording a strong instead of weak of A-B interaction with respect to that of sample sintered at 1100°C.

With respect to A-A interactions in the unit cell of $Ni_{0.5}Zn_{0.5}Fe_2O_4$, MEM results are consistent with the theoretical result (Lakhani *et al.*, 2011) for only in the case of samples sintered at 1000 and 1100°C. Weakening of A-A interactions in the samples sintered at 800, 900 and 1300°C and strengthening of A-A interactions in the samples sintered at 1200 and 1400°C are attributed to the presence of disorder in cation distribution in $Ni_{0.5}Zn_{0.5}Fe_2O_4$ ferrites unit cell. From the cation distribution (Table 2.2) of $Ni_{0.5}Zn_{0.5}Fe_2O_4$ samples it is observed that up to the sintered temperature of 1000°C, zinc atoms mostly occupy octahedral sites, due to Fe^{3+} ions migrating to A-sites. A greater number of Fe^{3+} ions at A-sites strengthen A-A interactions (Nasrin *et al.*, 2014). For 700°C sample, the number of Fe^{3+} ions occupying A-sites is maximum hence A-A interaction is very strong. Another reason for the strong A-A interaction in 700°C sample is attributed to the slight difference in the values of calculated and observed Bohr magneton (Table 2.16). With increase in sintering temperature, the number of Fe^{3+} ions at A-sites decreases and hence A-A interaction starts to decrease up to 1000°C. The sample sintered at 1100°C has more number of Fe^{3+} ions at the A-site (Table 2.2), when compared to the sample sintered at 1200°C and hence, the sample sintered at 1100°C must have stronger A-A interaction, but on the contrary MEM result (Table 2.15) shows that the sample sintered at 1200°C has recorded stronger A-A interaction than the sample sintered at 1100°C. This contradictory can be explained by examining SEM images. SEM (Figs.2.7(e-f)) reveals the existence of pores in samples sintered at 1100 and 1200°C. The pores tend to hinder free movement of magnetic walls during the magnetization process. As a consequence, intensity of effective magnetic field applied to material is reduced (Satishverma *et al.*, 2013). The "Field at M_s" value (Table 2.16) in the case of sample sintered at 1100°C is less when compared with that of sample sintered at 1200°C, which leads to infer that the amount of porosity in the sample sintered at 1100°C is high, thus leading to a weak interactions between various sites in the unit cell. Porosity and presence of other phases (Singh *et al.*, 2010; Abdullah *et al.*, 2010; Sozeri *et al.*, 2012) are among the important factors that influences the performance of a magnetic material. Hence, A-A interaction of the sample sintered at 1200°C is stronger than that of the sample sintered at 1100°C. The presence of Yaffet-Kittel angle in the sample sintered at 1200°C also strengthens the A-A interaction.

Regarding the B-B interactions in the unit cell of $Ni_{0.5}Zn_{0.5}Fe_2O_4$ samples, MEM and theoretical results (Lakhani *et al.*, 2011) agreed only in the case of samples sintered at 1300 and 1400°C. The contradiction between the MEM result and theoretical conclusion is explained as follows. The observed strengthening of B-B interactions up to the sample sintered at 900°C is attributed to the decrease of zinc atoms at the B-site (Table 2.2). Weakening of B-B interaction in the case of 1000°C is attributed to the sudden decrease of nickel atoms at the B-site. The total occupation of zinc atoms at the B-site affects the B-B interaction of 1100°C sample.

MEM result and theoretical conclusion arrived by Lakhani et al. (Lakhani *et al.*, 2011) completely contradict for the sample sintered at 1200°C. The contradiction in the sample sintered at 1200°C is attributed to the (i) presence of a Yaffet-Kittel angle in 1200°C, though small in magnitude, due to which A-A interactions may get strengthened than that of 1100°C and (ii) presence of second phase, which is high in 1100°C among the studied samples, may weaken the interactions including A-A. Since, in the present study, the $Ni_{0.5}Zn_{0.5}Fe_2O_4$ samples differ from each other only by means of sintering temperature and not in composition, hence the magnetic parameters depend to a greater extent on grain size, grain boundary, structural defects, cation distribution, sintering temperature and presence of other phases (Singh *et al.*, 2010; Sozeri *et al.*, 2012). Due to small Yaffet-Kittel angle value, the weakening of A-B interactions is not very pronounced in the sample sintered at 1200°C. Although A-B interaction remain the strongest interactions in 1200°C, A-B and A-A interactions are also comparable in strength (Table 2.15).

3.4.3.2 $Ni_{0.8}Zn_{0.2}Fe_2O_4$ sample

MEM result shows that A-B interaction is the strongest interaction among the three (A-A, A-B, B-B) interactions in $Ni_{0.8}Zn_{0.2}Fe_2O_4$ sample. Absence of the Yaffet-Kittel angle (Table 2.16) also confirms the MEM result. From Table 2.15, it can be seen that A-A interaction is stronger than B-B interaction which is attributed to occupation of Fe^{3+} ions at A-sites (Nasrin *et al.*, 2014).

3.4.3.3 $Mn_{0.4}Zn_{0.6}Fe_2O_4$ and $Co_{0.4}Zn_{0.6}Fe_2O_4$ samples

A close examination of interionic distances and angles of $Mn_{0.4}Zn_{0.6}Fe_2O_4$ and $Co_{0.4}Zn_{0.6}Fe_2O_4$ samples (Table 2.13(b)), one would expect a weak A-B, A-A and a strong B-B interactions but on contrary the mid bond electron density values calculated using MEM show a strong A-B, B-B and A-A interactions for $Mn_{0.4}Zn_{0.6}Fe_2O_4$ than that of $Co_{0.4}Zn_{0.6}Fe_2O_4$ sample. This contradiction can be understood by taking into account the substitutions involved. In the theoretical assumption (Lakhani *et al.*, 2011), it was a series of substitutions of same element whereas in the present study the substitutions involved are different but with same composition. The magnetic moment per formula unit for Co is $3\mu_B$ whereas for Mn it is $5\mu_B$. Hence, comparatively a strong A-B, A-A and B-B interactions in $Mn_{0.4}Zn_{0.6}Fe_2O_4$ sample is observed. The A-A interactions are more stronger than B-B interactions in both samples which are attributed to the disorder in cation distribution and presence of additional phases (Nasrin *et al.*, 2014).

3.4.3.4 $Ni_{0.53}Cu_{0.12}Zn_{0.35}Fe_2O_4$ and $Mg_{0.2}Cu_{0.3}Zn_{0.5}Fe_2O_4$ samples

MEM result reveals that A-A interaction is the strongest interaction in $Ni_{0.53}Cu_{0.12}Zn_{0.35}Fe_2O_4$ and $Mg_{0.2}Cu_{0.3}Zn_{0.5}Fe_2O_4$ samples. $Ni_{0.53}Cu_{0.12}Zn_{0.35}Fe_2O_4$ sample has stronger A-A, A-B and B-B interactions than $Mg_{0.2}Cu_{0.3}Zn_{0.5}Fe_2O_4$ sample. The magnetic moment per formula unit for Ni is $2\mu_B$, Cu is $2\mu_B$, whereas for Zn and Mg it is $0\mu_B$, which results in a strong A-B, A-A and B-B interactions in $Ni_{0.53}Cu_{0.12}Zn_{0.35}Fe_2O_4$ sample. The A-A interactions are stronger than B-B interactions in both the samples which are attributed to the disorder in cation distribution and presence of additional phases (Nasrin *et al.*, 2014).

3.4.3.5 $NiFe_2O_4$ sample

From Table 2.15, it can be seen that A-A interaction is the strongest, followed by A-B and B-B interactions in $NiFe_2O_4$ sample. Presence of Ni^{2+} ions at the A site (Nasrin *et al.*, 2014) strengthen the A-A interactions in $NiFe_2O_4$.

3.5 Discussion on magnetic properties

The hysteresis loop of the synthesized samples was obtained at room temperature using a vibrating sample magnetometer and is presented in Figs.2.16(a-f). Various magnetic properties such as, saturation magnetization (M_s), field at M_s, coercivity (H_c), retentivity (M_r), squareness ratio (M_r/M_s) and magnetic moment per formula unit in Bohr magneton (μ_B) obtained from both hysteresis loop (μ_B^H) and cation distribution, (μ_B^N) are tabulated in Table 2.16.

3.5.1 Saturation magnetization of the synthesized samples from the hysteresis curve

The hysteresis curve of $Ni_{0.5}Zn_{0.5}Fe_2O_4$ samples reveals that the magnetization increases and attains saturation with increase in applied field. The saturation magnetization (M_s) increases with increase in sintering temperature except for the samples 1100, 1200 and 1400°C samples (Table 2.16). The variation in M_s value with sintering temperature is attributed to distribution of Fe^{3+} ions over the available A and B site. According to Neel's two sub lattice model (Abdullah *et al.*, 2010) the net magnetization is the difference between the magnetization A and B sites. The variation in M_s with the content of Fe^{3+} ion at A-site is shown in Fig.2.17.

Migration of Fe^{3+} ion from A-site to B-site causes the B-site magnetization to increase and that of A-site to decrease, resulting in increase in the net magnetization. In the present study, the M_s value of $Ni_{0.5}Zn_{0.5}Fe_2O_4$ samples are low when compared to the reported value of 73emu/g (Narayanaswamy *et al.*, 2008) except the samples sintered at 1000, 1300 and 1400°C. Since the composition remains the same, the obtained low values of M_s in the present study is attributed to the combined effect of cation distribution, presence of second phase, nature of domain state and lattice parameter value (Singh *et al.*, 2010; Sozeri *et al.*, 2012). Low value of M_s in the case of 700 and 1100°C samples can be explained as a combination of (i) higher lattice parameter (Naughton *et al.*, 2007) and (ii) migration of more Fe^{3+} ions from B to A sites and Zn^{2+} ions from A to B sites (Narayanaswamy *et al.*, 2008). Presence of Zn^{2+} ions in the octahedral B-site, forcing Fe^{3+} ions into tetrahedral A-site, causing the B-site magnetization and hence, the net magnetization to decrease. Among the samples of $Ni_{0.5}Zn_{0.5}Fe_2O_4$ series, the sample sintered at 1300°C is highly ferromagnetic and the sample sintered at 700°C sample shows paramagnetic behaviour.

The saturation magnetization value of $Ni_{0.8}Zn_{0.2}Fe_2O_4$ sample is 50.55emu/g (Fig2.16(c)). Magnetic saturation of about 64.03emu/g is reported by Jadhav et al. (Jadhav *et al.*, 2008) for this particular composition prepared by co-precipitation method. This decreased value of saturation magnetization in the present case is attributed to the migration of zinc ions to the B-site, forcing iron ions to the A-site. Thus, the magnetic moment of the B-site is decreased and that of the A-site is increased and hence, a decrease in the saturation magnetization. From the hysteresis curve, it can be seen that the $Ni_{0.8}Zn_{0.2}Fe_2O_4$ sample exhibits a ferromagnetic nature.

Ferromagnetic and paramagnetic nature of the $Co_{0.4}Zn_{0.6}Fe_2O_4$ and $Mn_{0.4}Zn_{0.6}Fe_2O_4$ samples (Fig.2.16(d)) is evidenced from the shape of the hysteresis loop. The saturation magnetization (M_s) obtained in the present study for $Co_{0.4}Zn_{0.6}Fe_2O_4$ is 18.62 emu/g whereas a value of 68 and 59 emu/g is reported by Hou et al. (Hou *et al.*, 2010) and Nikam et al. (Nikam *et al.*, 2015) respectively. The M_s value of $Mn_{0.4}Zn_{0.6}Fe_2O_4$ sample in the present study is 6.66 emu/g whereas the literature survey reports 48 emu/g (Dasgupta *et al.*, 2006) and 30 emu/g (Mirshekari *et al.*, 2012). Low M_s value in both samples in the present study is attributed to occupation of zinc at the B site (Table 2.16) due to which the B-site magnetization decreases while that of A increases hence, decrease in net magnetization. Crystalline nature and presence of secondary phases are also the reasons for a low saturation magnetization value (Singh *et al.*, 2010; Sozeri *et al.*, 2012).

The magnetization increases and attains saturation with increase in applied field in the case of $Ni_{0.53}Cu_{0.12}Zn_{0.35}Fe_2O_4$ whereas $Mg_{0.2}Cu_{0.3}Zn_{0.5}Fe_2O_4$ (Fig.2.16(e)) does not attain saturation even at 20kG. The shape of hysteresis loop of $Ni_{0.53}Cu_{0.12}Zn_{0.35}Fe_2O_4$ indicates the ferromagnetic nature of the sample whereas that of the $Mg_{0.2}Cu_{0.3}Zn_{0.5}Fe_2O_4$ indicates super-paramagnetic nature which is supported by the negligible values of H_c, M_r and non saturation of M_s. The M_s values for bulk $NiFe_2O_4$ and $ZnFe_2O_4$ are 50 and 5 emu/g respectively (Tehrani *et al.*, 2012) whereas from the Table 2.16, it can be seen that the M_s values of $Ni_{0.53}Cu_{0.12}Zn_{0.35}Fe_2O_4$ and $Mg_{0.2}Cu_{0.3}Zn_{0.5}Fe_2O_4$ sample (58.03 and 11.58 emu/g respectively) are greater than that of bulk which are attributed to the increase in crystalline nature and density (Reddy *et al.*, 2012). The variation in the M_s value of $Ni_{0.53}Cu_{0.12}Zn_{0.35}Fe_2O_4$ and $Mg_{0.2}Cu_{0.3}Zn_{0.5}Fe_2O_4$ is obvious, since the octahedral site of $Ni_{0.53}Cu_{0.12}Zn_{0.35}Fe_2O_4$ is occupied by 13% of non-magnetic (Zn) ions whereas in $Mg_{0.2}Cu_{0.3}Zn_{0.5}Fe_2O_4$ it is (Zn and Mg) almost 25% (Naseri *et al.*, 2011).

The 'S' shape hysteresis curve of $NiFe_2O_4$ sample (Fig.2.16(f)), together with small coercivity value, reveals the presence of small magnetic particles exhibiting super paramagnetic behavior. The saturation magnetization (M_s) value of $NiFe_2O_4$ sample at 300K is 11.6 emu/g. It is much lower when compared with that of the bulk $NiFe_2O_4$ value, 55 emu/g (Goldman, 2006) and it is attributed to (i) the core shell morphology of the nano particles consisting of ferrimagnetically aligned core spins and spin-glass-like surface layer (Nabiyouni et al., 2010) and (ii) the lower degree of crystallinity of the sample (Azadmanjiri et al., 2007). The presence of spin glass shell tends to decrease the number of aligned magnetic moment and this fact is reflected in the discrepancies between the calculated and observed value of Bohr magneton. The intensity of the (311) peak indicates the lower degree crystalline nature of the sample.

3.5.2 Coercivity of the synthesized samples from the hysteresis curve

The increase in sintering temperature causes the grain size (Table 2.8) to increase which in turn decrease coercivity (H_c) value of $Ni_{0.5}Zn_{0.5}Fe_2O_4$ samples as shown in Fig.2.18.

This decrease in coercivity value is attributed to the increase in number of walls with the increase in grain size. Critical size (D_{cr}), from single domain to multi domain in the spherical particle model, is calculated using Eq.(1.56). T_C = 550 K, M_s = 310 G, K_1 = 1.5×10^3 J/m^3 and a = 8.4Å (Gao *et al.*, 2013) were substituted in Eq.(1.56) and the critical diameter value is 24nm which is consistent with the result of 17(2) nm to 34(3) nm for the samples sintered from 700°C to 1400°C in the present study (Table 2.8). Hence coercivity value of the samples starts decreasing once the average grain size (t) is greater than the critical diameter size (D_{cr}). For the sample sintered at 700°C, 't' is less than D_{cr} whereas for other samples 't' is greater than D_{cr} which means the particles in 800°C to 1400°C are in multi domain state while it is single domain in 700°C. Thus in single domain nanocrystallites surface effect of the particles becomes important, due to which 700°C sample has low M_s value (Table 2.16).

Particles in multi domain state ('t' greater than 'D_{cr}') will require fewer magnetic fields for domain wall rotation (Zhang *et al.*, 2013) which is confirmed by the experimental value of "Field at M_s" value (Table 2.16). Also the coercivity decreases with increasing sintering temperature (Table 2.16) except for the sample sintered at 1200°C. Coercivity value (Table 2.16) of the sample sintered at 1200°C is high when compared with that of the sample sintered at 1100°C indicates that the sample sintered at 1200°C belongs to the category of hard ferrite material. Grain size of the sample sintered at 1200°C is less when compared with that of the sample sintered at 1100°C and the presence of Yaffet-Kittel angle in 1200°C sample are attributed to the higher value of "Field at M_s" for 1200°C sample. The low squareness value (M_r / M_s) obtained, indicates the existence of significant amount of superparamagnetic particles fast relaxing at room temperature, when the external magnetic field is turned off (Maqsood *et al.*, 2012).

For $Ni_{0.8}Zn_{0.2}Fe_2O_4$ nano ferrites, the value for T_C is 593 K (Ramakrishna *et al.*, 2012). The value of D_{cr} is found out to be 25nm whereas the 't' value in the present study is 30(2) nm indicating that the particles are in multi domain state and the same is confirmed by the small value of M_r/M_s and could be useful as permanent magnets and in applications such as recording media of high density.

The magnetocrystalline anisotropy constant ($|K_1|$) value of Co (K_1= 2E5 J/m^3) is more than that of Mn (K_1= 3E3 J/m^3) and hence, the coercivity value of $Co_{0.4}Zn_{0.6}Fe_2O_4$ is more than that of $Mn_{0.4}Zn_{0.6}Fe_2O_4$. Presence of hematite impurity Fe_2O_3 phase is also contributing to the increase in coercivity value (Rashad *et al.*, 2012) since, Fe_2O_3 has high anisotropy constant value of the order of K_1= 1.4E4 J/m^3 (Chen *et al.*, 2013).

Critical size (D_{cr}) of $Ni_{0.53}Cu_{0.12}Zn_{0.35}Fe_2O_4$ (Sujatha *et al.*, 2013) is 39 nm whereas the same for $Mg_{0.2}Cu_{0.3}Zn_{0.5}Fe_2O_4$ is (Bhosale *et al.*, 1999) is 14 nm. In the present study, the grain size of $Ni_{0.53}Cu_{0.12}Zn_{0.35}Fe_2O_4$ and $Mg_{0.2}Cu_{0.3}Zn_{0.5}Fe_2O_4$ is evaluated as 29(6) and 34(3) nm respectively. Thus $Ni_{0.53}Cu_{0.12}Zn_{0.35}Fe_2O_4$ is in single domain state whereas $Mg_{0.2}Cu_{0.3}Zn_{0.5}Fe_2O_4$ is in multi domain state. The magnetocrystalline anisotropy $(|K_1|)$ constant value of $Ni_{0.53}Cu_{0.12}Zn_{0.35}Fe_2O_4$ (3.65E3 J/m^3) (Sujatha et al., 2013) is larger than that of $Mg_{0.2}Cu_{0.3}Zn_{0.5}Fe_2O_4$ (2.6E3 J/m^3) sample (Rashad et al., 2012). Hence, in the present study the obtained coercivity value is attributed to the combined effect of Fe_2O_3 phase, K_1 and D_{cr} values. The Permeability (μ) value suggests that magnetization of $Ni_{0.53}Cu_{0.12}Zn_{0.35}Fe_2O_4$ is more than that of $Mg_{0.2}Cu_{0.3}Zn_{0.5}Fe_2O_4$ sample.

The squareness value of M_r/M_s (less than 0.5) of all samples suggests that both samples could be used to obtain core and coils with low inductance (Druc et al., 2013).

Pulišová et al. (Pulišová et al., 2013) reported that D_{cr} for $NiFe_2O_4$ nano particles is 100 nm whereas the particle size in the present is 9(1) nm indicating that the synthesized nano particles is in single domain state which is further evidenced by high value of "Field at M_s" (Table 2.16).

3.5.3 Evaluation of magnetic moment from saturation magnetization and cation distribution data

The magnetic moment per formula unit in units of Bohr magneton (μ_B) obtained from the hysteresis data (μ_B^H) and from the knowledge of cation distribution, known as Neel's magnetic moment (μ_B^N) is evaluated using Eq.1.57 and Eq.1.58 respectively for all the synthesized samples and are tabulated in Table 2.16.

For $Ni_{0.5}Zn_{0.5}Fe_2O_4$ samples, the agreement between μ_B^H and μ_B^N justifies the cation distributions and confirmed the collinear spin ordering, except the samples sintered at 700 and 1200°C. Disagreement between μ_B^H and μ_B^N indicates existence of Yaffet-Kittel (YK) angle which is evaluated using Eq.1.59. is tabulated in Table 2.16. MEM study also reveals the presence of Yaffet-Kittel angle in the samples sintered at 700 and 1200°C by means of recording the strongest A-A interaction in the sample sintered at 700°C and in the case of sample sintered at 1200°C, the strength of A-A and A-B interactions are almost comparable.

The increase in magnetic moment per formula unit in units of Bohr magneton (μ_B^H) with the increase in sintering temperature up to 1000°C can be attributed to migration of Fe^{3+} ions from A to B site (Choodamani et al., 2013) and thus indicates an increase in ferromagnetic behaviour (Ramakrishna et al., 2012). The permeability (μ) increases due to the increase in grain size. An increase in the density raises the spin rotational contribution which in turn increases the permeability (Shrotri et al., 1999).

A sharp decrease in μ_B^H is observed in the case of sample sintered at 1100°C and thereafter μ_B^H increases with increase in sintering temperature, though a slight decrease in μ_B^H is observed in sample sintered at 1400°C. The decrease in μ_B^H is attributed to the occupation of zinc atoms at B-site leading to decrease in ferromagnetic nature of the sample as evidenced from Fig. 2.16(a), with the decrease in μ_B^H. The values of saturation magnetization and Bohr magneton are larger in the case of sample sintered at 1300°C which means the ferromagnetic nature is high and the same is confirmed from the MEM studies by exhibiting strongest A-B interaction among the samples sintered at different temperatures from 700 to 1400°C. The samples sintered at 1300 and 1400°C exhibit soft ferrite material characters with high saturation magnetization (M_s) and low coercivity (H_{ci}) values.

The agreement between μ_B^H and μ_B^N (Table 2.16) justifies the cation arrangement in the case of $Ni_{0.8}Zn_{0.2}Fe_2O_4$ sample.

The disagreement between μ_B^H and μ_B^N in the case of $Mn_{0.4}Zn_{0.6}Fe_2O_4$ and $Co_{0.4}Zn_{0.6}Fe_2O_4$ samples (Table 2.16) indicates the presence of Yaffet-Kittel (YK) angle (α_{yk}). The presence of YK angle means collinear spin ordering is not possible at A and B sites in both samples, which will lead to decrease in magnetization at B-site which in turn decrease the net magnetization. Hence, low saturation magnetization value in both samples is due to the presence of YK angle besides cation disorder and presence of second phases (Singh et al., 2010; Sozeri et al., 2012). MEM and magnetic study results in the present study completely contradicted to each other. As per the MEM study (Table 2.15) A-B interactions is the strongest interactions among the A-A, A-B and B-B interactions, which indicates the existence of collinear spins at the A and B sublattice whereas the magnetic study reveals the existence of YK angle from which one can conclude that collinear spins is not possible at the 'B' sublattices. This contradicted result is attributed to contraction in the experimental lattice parameter a_{exp} value. Naughton et al. (Naughton et al., 2007) have reported that any expansion in lattice parameter would weaken the A-B super exchange interaction. An alteration in the lattice spacing would alter the overlapping of the 3d and 2p wave function on adjacent metal and oxygen ions. The a_{exp} value of both samples is smaller than that of undoped zinc ferrite (8.441Å) and hence comparatively enhancing the overlap of the 3d and 2p wave function leading to strengthening of A-B interactions in both samples. Increase in the values of magnetic moment per formula unit in Bohr magneton and in the permeability of $Co_{0.4}Zn_{0.6}Fe_2O_4$ when compared to $Mn_{0.4}Zn_{0.6}Fe_2O_4$ confirms the ferromagnetic nature of $Co_{0.4}Zn_{0.6}Fe_2O_4$ (Ramakrishna et al., 2012).

The disagreement between μ_B^H and μ_B^N in the case of $Ni_{0.53}Cu_{0.12}Zn_{0.35}Fe_2O_4$ and $Mg_{0.2}Cu_{0.3}Zn_{0.5}Fe_2O_4$ (Table 2.16) samples indicate the presence of Yaffet-Kittel (YK) angle. The presence of Yaffet-Kittel angle means collinear spin ordering is not possible at A and B sites in both samples. Results from MEM in the present study completely contradict the results from magnetic studies. This contradicted result may be explained as follows. It is well known in the literature (Naseri *et al.*, 2011) that any disorder in cation distribution leads to a large degree of inversion (occupation of Fe^{3+} ions at the A site) and occurrence of strong A-B interactions in smaller size particles than in larger size particles. Hence the saturation magnetization increases for smaller size particles, hence strengthening of A-B interactions may occur in the present study. The high magnetic moment per formula unit in Bohr magneton of $Ni_{0.53}Cu_{0.12}Zn_{0.35}Fe_2O_4$ indicates ferromagnetic behaviour (Ramakrishna *et al.*, 2012) which is also supported by high permeability value of $Ni_{0.53}Cu_{0.12}Zn_{0.35}Fe_2O_4$ sample.

Due to the discrepancy between μ_B^H and μ_B^N values of Bohr magneton, Neel two sub-lattice model cannot be used in $NiFe_2O_4$ sample. Hence, the Yaffet-Kittel (θ_{YK}) angle is calculated. According to the Yaffet-Kittel model, the B sublattice can be split into two sublattices B_1 and B_2 having moments equal in magnitude and each making an angle (θ_{YK}) with the direction of the net magnetization at 0K. This triangular spin arrangement of ions on B site leads to the reduction in the A – B interactions, which is also confirmed from the MEM study (Table 2.15), hence low saturation magnetization value is observed.

3.6 Discussion on optical properties

The optical band gap energy (E_g) of the synthesized samples is estimated using Eq.1.60 and are tabulated in Table 2.17. The variation of band gap energy of $Ni_{0.5}Zn_{0.5}Fe_2O_4$ samples with sintering temperature is shown in Figs. 2.19(a-b). The direct band gap energy values of the samples 700 to 1400°C respectively are 2.02, 2.00, 1.98, 1.97, 2.03, 2.12, 2.15, and 2.20 eV respectively. It can be seen that E_g value decrease slightly upto 1000°C and thereafter increases. A band gap energy of 2.56 eV is reported for the $Ni_{0.5}Zn_{0.5}Fe_2O_4$ nano ferrite samples prepared by sol-gel process (Tehrani *et al.*, 2012) whereas a band gap energy of 1.70 eV is being reported for $Ni_{0.5}Zn_{0.5}Fe_2O_4$ nano ferrite pigments prepared by sol-gel auto combustion method (Sutka *et al.*, 2012). The band gap energy value is affected by various factors such as lattice parmaters, crystallite size and presence of impurities (Seema *et al.*, 2014). The decrease in E_g with sintering temperature upto 1000°C is attributed to the increase in particle size and decrease in lattice parameter (a_{exp}) of the samples (Salahudeen *et al.*, 2014) whereas the effect of impurities is pronounced and thereby influencing the E_g value of the samples sintered from 1100 to 1400°C of $Ni_{0.5}Zn_{0.5}Fe_2O_4$.

The E_g value of $Ni_{0.8}Zn_{0.2}Fe_2O_4$ sample is 2.29 eV (Fig.2.19(c)). Gao et al. (Gao *et al.*,2010) reported E_g value of 1.9eV for $ZnFe_2O_4$ nanocrystalline prepared by polymer complex method whereas E_g value of 2.5eV is reported for $NiFe_2O_4$ prepared by co-precipitation method (Dolia *et al.*, 2006).

In the case of $Mn_{0.4}Zn_{0.6}Fe_2O_4$ and $Co_{0.4}Zn_{0.6}Fe_2O_4$ samples, the E_g values are found to be 2.17eV and 2.51eV respectively (Fig.2.19(d)). Demir et al. (Demir *et al.*, 2014) and Hema et al. (Hema *et al.*, 2015) have reported direct band gap energy values of $Mn_{0.4}Zn_{0.6}Fe_2O_4$ and $Co_{0.4}Zn_{0.6}Fe_2O_4$ respectively as 1.88eV and 2.51eV. The increase in the band gap energy value for $Mn_{0.4}Zn_{0.6}Fe_2O_4$ sample is attributed to the presence of impurities and additional phases (Pathak *et al.*, 2013).

For $Ni_{0.53}Cu_{0.12}Zn_{0.35}Fe_2O_4$ and $Mg_{0.2}Cu_{0.3}Zn_{0.5}Fe_2O_4$, the band gap energy E_g is 2.04 and 2.76 eV respectively (Fig.2.19(e)) and it is attributed to the combined effect of lattice parameter, particle size values and to the presence of additional phase in the samples.

In the case of $NiFe_2O_4$, E_g is found out to be 2.1eV (Fig.2.19(f)) which is comparable with the band gap energy value of 2.19eV of nickel ferrite prepared by novel combustion synthetic method (Balaji *et al.*, 2005).

3.7 Discussion on dielectric properties

3.7.1 Variation of dielectric constant with frequency of $Ni_{0.5}Zn_{0.5}Fe_2O_4$ samples

The effect of sintering temperature on real (ε') and imaginary (ε'') part of dielectric constant of the samples measured at room temperature in the frequency range from 0.1Hz to 15MHz is shown in Figs.2.20(a and b) respectively and the various dielectric parameters of $Ni_{0.5}Zn_{0.5}Fe_2O_4$ samples sintered at different temperatures are tabulated in Table 2.18. All the sintered samples under investigation have recorded normal dielectric behavior. From Figs.2.20(a and b), it can be seen that the sample sintered at 700°C shows maximum dielectric dispersion in ε' and ε'', followed by sample sintered at 900°C. This dispersion in ferrite, according to Maxwell-Wagner model (Maxwell, 1973; Wagner, 1913) and Koop's phenomenological theory (Koops, 1951) of dielectrics can be attributed to different degrees of conduction by grain (good conductor) and grain boundaries (poor conductor) of ferrite. At lower frequencies, the grain boundaries are more effective in electrical conduction than grains. The dielectric constant ε' value, at the lowest studied frequency, decreases with increase in sintering temperature except for the sample sintered at 800°C and 1200°C. The samples sintered at 700°C and 900°C show ε' value of 16620 and 3466 respectively whereas the value of ε' of all other sample lies between 2000 and 500.

High ε' value in a sample can be attributed to (i) thickness of grain boundary (Meaz et al., 2005) (ii) presence of more number of Fe^{2+} ions available on octahedral (B) site (Tehrani et al., 2012) and (iii) presence of minute secondary phase (Abdullah et al., 2010). More number of Fe^{2+} ions on B-sites should lead to high dielectric loss (Rangamohan et al., 1999) but the sample sintered at 700°C shows the least dielectric loss (Fig.2.20c) and hence, the effect of thickness of the grain boundary and the presence of second phase are more pronounced than the effect of presence of higher number of Fe^{2+} ions on B-sites.

The value of ε' for the sample sintered at 900°C is more than that of the sample sintered at 800°C and this could be explained on the basis of the combined effect of changes in the value of lattice parameter 'a_{exp}' (Table 2.6) and octahedral bond length 'p' (Table 2.13a). The octahedral bond length 'p' of the sample sintered at 900°C is larger than that of the sample sintered at 800°C and hence the dipole moment and consequently the polarization is more for the sample sintered at 900°C. The lattice parameter 'a_{exp}' of the sample sintered at 900°C is less than that of the sample sintered at 800°C hence decrease in unit cell volume. This increases the dielectric constant value of the sample sintered at 900°C than the sample sintered at 800°C (Nabiyouni et al., 2010; Bhatacharjee et al., 2011).

Similarly, the dielectric constant ε' of the sample sintered at 1200°C is more than that of the samples sintered at 1000 and 1100°C and it is attributed to the lower porosity and higher density of 1200°C sample (Mangalaraja et al., 2004). Microstructure image of the samples is shown in Figs.2.7(a-f) to visualize the variation in grain size with sintering temperature.

Slightly higher ε' value of the sample sintered at 1000°C than the sample sintered at 1100°C is attributed to the presence of more Fe^{2+} ions at B site, which is reflected in the radii of ions present at tetrahedral and octahedral ($r_{A(XRD)}$, $r_{B(XRD)}$) sites (Kannan et al., 2017; Saravanan et al., 2017).

The imaginary part of dielectric constant (ε'') also decreases with increase in sintering temperature at the lower frequencies, except for the sample sintered at 1200°C.

3.7.2 Variation of tanδ with frequency of $Ni_{0.5}Zn_{0.5}Fe_2O_4$ samples

Fig.2.20(c) shows the variation of tanδ with frequency, measured at room temperature for the samples sintered from 700°C to 1200°C and it can be seen that, at lower frequencies the dielectric loss decreases with decrease in sintering temperature except the sample sintered at 800°C. For the sample sintered at 700°C, the tanδ value first increases, reaches maximum and then decreases with the increase in frequency. For the samples sintered at 900, 1000 and 1100°C, the tanδ curve decreases with the increase in frequency, leading to assume that these sample may also show maxima below the frequency studied in the present investigation. For the samples sintered at 800 and 1200°C, presence of peak at the lowest studied frequency can be seen. Also, all the samples in the present study show a small peak (around log f = 5.5) in the higher frequency range.

In ferrite materials the dielectric loss with frequency is ascribed to the presence of structural defects like impurities and imperfections in crystal lattice and also to the lag of the polarization behind the applied alternating electric field (Verma et al., 1999). Also the presence of peak at a particular frequency indicates resonance occurring between the hopping frequency of electrons in different valance states (Fe^{2+}, Fe^{3+}) becomes equal to the frequency of applied field (Abdullah et al., 2010).

3.7.3 Variation of AC conductivity with frequency of $Ni_{0.5}Zn_{0.5}Fe_2O_4$ samples

The variation of ac conductivity with frequency at room temperature of the samples under investigation is shown in Fig.2.20(d). The conduction mechanism is due to hopping of charge carriers between Fe^{2+} and Fe^{3+} ions on octahedral sites. With increase in sintering temperature, the conductivity decreases which can be attributed to the decrease in number of Fe^{2+} ions on octahedral sites.

With the increase in frequency, all sintered samples show increase in conductivity values, which are termed as normal behavior of ferrites. This behavior is due to the enhancement of hopping of charge carriers on octahedral sites, resulting in an increase in the conducting process thereby increasing the conductivity. The presence of small amount of Fe^{2+} ions, which plays dominant role in the conduction mechanism, depends upon the concentration of Fe^{2+}/Fe^{3+} ions present on octahedral site. The Fe^{2+} ions concentration at octahedral site depends upon parameters like sintering temperature, cooling rate, type of annealing etc. (Trivedi et al., 2005).

The sample sintered at 700°C shows the highest value of ε' and σ_{ac} in the present study and it is attributed to the presence of more number of Fe^{2+} ions at octahedral site. The sample sintered at 1100°C shows the least value of ε' and σ_{ac} in the present study and it is attributed to the presence of less number of Fe^{2+} ions at octahedral site. The increase in density value and the grain size also contributed to the decrease in conductivity with sintering temperature (Rao et al., 1997).

The thickness and diameter of $Ni_{0.5}Zn_{0.5}Fe_2O_4$ and $NiFe_2O_4$ samples are tabulated in Table 2.19.

3.7.4 Variation of dielectric constant with frequency of $NiFe_2O_4$ samples

Fig.2.21(a) shows the real and imaginary part of dielectric constant behavior of $NiFe_2O_4$ samples. The fall in the dielectric constant value is steep in the lower frequency region whereas in the higher frequency region almost constant value of dielectric constant is observed. Normal dielectric behaviour is observed since the value of dielectric constant decreases with increase in frequency without showing any peak. Several investigators (Rangamohan et al., 1999; Trivedi et al., 2005) also reported normal dielectric behaviour in ferrites. This normal dielectric behavior can be explained on the basis of Maxwell-Wagner interfacial polarization (Maxwell, 1973; Wagner, 1913) and Koop's phenomenological theory (Koops, 1951). In ferrites, the system may be thought of as consisting of well conducting grains separated by highly resistive grain boundaries. The charge carriers upon reaching the grain boundary by hopping, encounter resistance so that accumulation of charge at the boundaries causing interfacial polarization and hence the dielectric constant value is highly raised. A hopping electron in the direction of the external applied field increases the dielectric polarization, therefore, there is a strong correlation between the conduction mechanism and the dielectric behavior in ferrites. However, as the frequency of the applied field is increased, the dielectric constant decreases until reaching a constant value, because beyond a certain frequency of the applied electric field, the polarization cannot keep pace with the applied field (ElKony et al., 2012). In the present study, dielectric constant value of 10^6 and 11 respectively is observed at the lowest (0.1Hz) and highest frequency studied (15MHz). This high value (10^6) indicates that the available number of Fe^{2+} ions at B site is more. These ferrous ions on B sites, which takes part in the electron exchange $Fe^{2+} \leftrightarrows Fe^{3+}$ is responsible for the polarization and hence dielectric constant in ferrites. Ferrites with high value (10^6) of dielectric constant find applications in microwave devices (Ramanareddy et al., 1999). Rangamohan et al. (Rangamohan et al., 1999) observed a value of ε' around 32 at 100 kHz for nickel ferrite sample prepared by double sintering ceramic technique whereas in the present work a value of ε' about 34 is observed at 100 kHz. From the Fig.2.21(a) it can be seen that both ε' and ε'' follows the same trend with increase in the frequency.

The plot of dielectric loss tangent (tanδ) against log frequency is shown in Fig2.21(b). It is observed that initially, the tangent loss increases, reaches a maximum and then decreases with the increase in frequency. In $NiFe_2O_4$ sample, tanδ shows a maximum at a frequency of 6 Hz. The occurrence of the peak in the plot tanδ versus log frequency curve for $NiFe_2O_4$ sample is explained (Rangamohan et al., 1999). Accoeding to Iwauchi (Iwauchi, 1971) there is a strong correlation between the conduction mechanism and the dielectric behavior of ferrites. The conduction mechanism in n-type ferrites is considered to be due to the hopping of electrons between Fe^{2+} and Fe^{3+} and the hopping of holes between Ni^{3+} and Ni^{2+} in p-type ferrites. As such, when the hopping frequency is nearly equal to that of the externally applied electric field, a maximum of tangent loss may be observed. The dielectric parameter (ε', ε'' and tanδ) values at selected frequencies are listed in Table 2.20.

References

[1]   Abdullah DM, Khalid MB, Vivek V, Siddiqui WA, Kotnala RK (2010) J Alloys Compd 493:553-560. https://doi.org/10.1016/j.jallcom.2009.12.154

[2]   Azadmanjiri J, Seyyed Ebrahimi SA, Salehani HK (2007) Ceram Int 33:1623-1625. https://doi.org/10.1016/j.ceramint.2006.05.007

[3]   Azaroff LV (2009) Introduction to solids TMH Edition, Tata McGraw Hill Education Private Limited New Delhi.

[4]   Balaji S, Kalai Selvan R, Berchmans LJ, Angappan S, Subramanian K, Augustin CO (2005) Mater Sci Eng B 119:119-124. https://doi.org/10.1016/j.mseb.2005.01.021

[5]   Barati MR (2009) J Alloy Compd 478:375-380. https://doi.org/10.1016/j.jallcom.2008.11.022

[6]   Batoo KM, Ansari MS (2012) Nanoscale Res Lett 7(112):1-14.

[7]   Bhattacharjee K, Ghosh CK, Mitra MK, Das GC, Mukherjee S, Chattopadhyay KK, (2011) J Nanopart Res 13:739-750. https://doi.org/10.1007/s11051-010-0074-4

[8]   Bhatu SS, Lakhani VK, Tanna AR, Vasoya NH, Buch JU, Sharma PU, Trivedi TN, Joshi HH, Modi KB (2007) Ind J Pure Appl Phys 45:596-608.

[9]   Bhosale DN, Verenkar VMS, Rane KS, Bakare PP, Sawant SR (1999) Mater Chem Phys 59:57-62. https://doi.org/10.1016/S0254-0584(99)00028-0

[10] Chen R, Christiansen MG, Anikeeva P (2013) ACS Nano 7(10):8990-9000. https://doi.org/10.1021/nn4035266

[11] Chhaya UV, Mistry BV, Bhavsar KH, Gadhvi MR, Lakhani VK, Modi KB, Joshi US (2011) Ind J Pure Appl Phys 49: 833-840.

[12] Choodamani C, Nagabhushana GP, Ashoka S, Prasad BD, Rudraswamy B, G.T. Chandrappa GT (2013) J Alloys Compd 578:103-109. https://doi.org/10.1016/j.jallcom.2013.04.152

[13] Dasgupta S, Kim KB, Ellrich J, Eckert J, Manna I (2006) J Alloys Compd 424:13-20. https://doi.org/10.1016/j.jallcom.2005.12.078

[14] Demir A, Sadik G, Yakup B, Sinem E, Abdülhadi B (2014) J Inorg Organomet Polym 24:729-736. https://doi.org/10.1007/s10904-014-0032-1

[15] Dolia SN, Rakesh S, Sharma MP, Saxena NS (2006) Ind J Pure and Appl Phys 44:774-776.

[16] Druc AC, Dumitrescu AM, Borhan AI, Nica V, Iordan AR, Palamaru MN (2013) Cent Eur J Chem 11(8):1330-1342.

[17] ElKony D, Saafan SA (2012) J Amer Sci 8(10):51-57.

[18] Gabal MA, Al-Thabaiti SA, El-Mossalamy EH, Mokhtar M (2010) Ceram Int 36:1339-1346. https://doi.org/10.1016/j.ceramint.2010.01.021

[19] Gagankumar, Jyoti Shah, Kotnala RK, Virender Pratap Singh, Sarveena, Godawari Garg, Sagar E. Shirsath, Khalid M. Batoo, Mahavir Singh (2015) Mater Res Bull 63:216-225. https://doi.org/10.1016/j.materresbull.2014.12.009

[20] Gangatharan S, Venkatraju C, Sivakumar K (2010) Mater Sci Appl 1:19-24.

[21] Gao D, Shi Z, Xu Y, Zhang J, Yang G, Zhang J, Wang X, Xue D (2010) Nanoscale Res Lett 5:1289 -1294. https://doi.org/10.1007/s11671-010-9640-z

[22] Gao P, Hua X, Degirmenci V, Rooney D, Khraisheh M, Pollard R, Bowmen RM, Rebrov EV (2013) J Magn Magn Mater 348:44-50. https://doi.org/10.1016/j.jmmm.2013.07.060

[23] Goldman A (2006) Modern Ferrites Technology (New York Springer) Chap.2.

[24] Guinier A (1994) X-ray Diffraction Dover Publications New York.

[25] Gull SF, Daniel GJ (1978) Nature 272:686-690. https://doi.org/10.1038/272686a0

[26] Haralkar SJ, Kadam RK, More SS, Sagar SE, Mane ML, Swati P, Mane DR (2013) Mater Res Bull 48:1189-1196. https://doi.org/10.1016/j.materresbull.2012.12.018

[27] Hema E, Manikandan A, Karthika P, Arulantony S, Venkatraman BR (2015) J Supercond Nov Magn 28:2539-2552. https://doi.org/10.1007/s10948-015-3054-1

[28] Hou C, Yu H, Zhang Q, Li Y, Wang H (2010) J Alloys Compd 491:431-435. https://doi.org/10.1016/j.jallcom.2009.10.217

[29] Israel S, Saravanakumar S, Renuertson R, Sheeba RAJAR, R. Saravanan R (2012) Bull Mater Sci 35:111-122. https://doi.org/10.1007/s12034-011-0264-4

[30] Iwauchi K (1971) Japan J Appl Phys 10:1520-1528. https://doi.org/10.1143/JJAP.10.1520

[31] Izumi F, Dilanian RA (2002) Recent research developments in Physics part II Vol.3 Transworld Research Network, Trivandrum.

[32]   Jadhav SS, Shirsath SE, Toksha BG, Shukla SJ, Jadhav KM (2008) Chin J Chem Phys 21:381-386. https://doi.org/10.1088/1674-0068/21/04/381-386

[33]   Jeyadevan B, Tohji K, Nkatsuka K (1994) J Appl Phys 76: 6325-6327. https://doi.org/10.1063/1.358255

[34]   Joshi GP, Saxena NS, Mangal R, Mishra A, Sharma TP, (2003) Bull Mater Sci 26(4):387- 389. https://doi.org/10.1007/BF02711181

[35]   Kannan YB, Saravanan R, Srinivasan N, Ismail I (2017) J Magn Magn Mater 423:217-225. https://doi.org/10.1016/j.jmmm.2016.09.038

[36]   Kelm K, Mader M (2006) Z Naturfosch 61b: 665-671.

[37]   Koops CG (1951) Phys Rev 83:121-124. https://doi.org/10.1103/PhysRev.83.121

[38]   Kurmude DV, Barkule RS, Raut AV, Shengule DR, Jadhav KM (2014) J Supercond Nov Magn 27(2):547-553. https://doi.org/10.1007/s10948-013-2305-2

[39]   Lakhani VK, Pathak TK, Vasoya NH, Modi KB (2011) Solid State Sci 13: 539-547. https://doi.org/10.1016/j.solidstatesciences.2010.12.023

[40]   Mangalaraja RV, Ananthakumar S, Manohar P, Gnanam FD (2002) J Magn Magn Mater 253:56-64. https://doi.org/10.1016/S0304-8853(02)00413-4

[41]   Mangalaraja RV, Manohar P, Gnanam FD (2004) J Mater Sci 39:2037-2042. https://doi.org/10.1023/B:JMSC.0000017766.07079.80

[42]   Maqsood M, Faraz A (2012) J Supercond Nov Magn 25:1025-1033. https://doi.org/10.1007/s10948-011-1343-x

[43]   Maxwell JC (1973) A Treatise on Electricity and Magnetism, Oxford University Press, New York, Vol.1, pp.828.

[44]   Meaz TM, Attia SM, Abo El Ata AM (2003) J Magn Magn Mater 257:296-305. https://doi.org/10.1016/S0304-8853(02)01212-X

[45]   Mirshekari GR, Daee SS, Mohseni H, Torkian S, Ghasemi M, Ameriannejad M, Hoseinizade M, Pirnia M, Pourjafar D, Pourmahdavi M, Gheisari K (2012) Adv Mater Res 409:520-525. https://doi.org/10.4028/www.scientific.net/AMR.409.520

[46]   Momma K, Izumi F (2006) Comm Crystallogr Comput IUCr Newslett 7:106-119.

[47]   Nabiyouni G, Jafari MF, Mozafari M, Amighian J (2010) Chin Phys Lett 27:126401-4. https://doi.org/10.1088/0256-307X/27/12/126401

[48]   Narayanaswamy A, Sivakumar N (2008) Bull Mater Sci 31:373-380. https://doi.org/10.1007/s12034-008-0058-5

[49] Naseri MG, Saion EB, Hashim M, Shaari AH, Ahangar HA (2011) Solid State Commun 151:1031–1035. https://doi.org/10.1016/j.ssc.2011.04.018

[50] Nasrin S, Hoque SM, Chowdhury FZ, Hossen MM (2014) IOSR-JAP 6: 58-65. https://doi.org/10.9790/4861-06235865

[51] Naughton BT, Clarke DR (2007) J Am Ceram Soc 90:3541-3546. https://doi.org/10.1111/j.1551-2916.2007.01980.x

[52] Nikam DS, Jadhav SV, Khot VM, Bohara RA, Hong CK, Mali SS, Pawar SH (2015) RSC Adv 5:2338-2345. https://doi.org/10.1039/C4RA08342C

[53] Patange SM, Sagar E. Shirsath, Jangam GS, Lohar KS, Santosh SJ, Jadhav KM (2011) J Appl Phy 109: 053909(1-9).

[54] Pathak TK, Vasoya NH, Natarajan TS, Modi KB, Tayade RJ (2013) Mater Sci Forum 764:116-129. https://doi.org/10.4028/www.scientific.net/MSF.764.116

[55] Petřiček V Dušek M, Palatinus L (2006) JANA2006: The Crystallographic Computing System, Institute of Physics, Academy of Sciences of the Czech Republic Praha 2000.

[56] Pulišová P, Ková J, Voigt A, Raschman P (2013) J Magn Magn Mater (2013) 341:93-99. https://doi.org/10.1016/j.jmmm.2013.04.003

[57] Ramakrishna K, Ravinder D, Vijaya Kumar K, Abraham Lincon Ch (2012) World J Condens Matter Phys 2:153-159.

[58] Ramakrishna K, Vijaya Kumar K, Ravinder D (2012) Adv Mater Phys Chem 2:185-191.

[59] Ramanareddy AV, Rangamohan G, Ravinder D, Boyanov BS, (1999) J Mater Sci 34: 3169-3176. https://doi.org/10.1023/A:1004625721864

[60] Rangamohan G, Ravinder D, Ramanareddy AV, Boyanov BS (1999) Mater Lett 40:39-45. https://doi.org/10.1016/S0167-577X(99)00046-4

[61] Rao BP, Rao KH (1997) J Mater Sci 32:6049-6054. https://doi.org/10.1023/A:1018683615616

[62] Rashad MM, Mohamed RM, Ibrahim MA, Ismail LFM, Abdel-Aal EA (2012) Adv Powder Tech 23: 315-323. https://doi.org/10.1016/j.apt.2011.04.005

[63] Rath C, Mishra NC, Anand S, Das RP, Sahu KK, Upadhyay C, Verma HC (2000) Appl Phys Lett 76:475-477. https://doi.org/10.1063/1.125792

[64] Reddy MP, Kim IG, Yoo DS, Madhuri W, Reddy NR, SivaKumar KV, Reddy RR (2012) Mater Sci Appl 3(3):628-632.

[65] Rietveld HM (1969) J Appl Crystallogr 2:65-71. https://doi.org/10.1107/S0021889869006558

[66] Salahudeen AG, Elias S, Abdul HS, Mazliana AK, Naif M, Al-Hada, Alireza (2014) J Nanomater 10.1155/2014/416765. https://doi.org/10.1155/2014/416765

[67] Santosh SJ, Sagar ES, Toksha BG, Shengule DR, Jadhav KM (2008) J Optoelectron Adv Mater 10:2644-2648.

[68] Santosh SJ, Sagar ES, Toksha BG, Shukla SJ, Jadhav KM (2008) Chin J Chem Phys 21(4) 381-386. https://doi.org/10.1088/1674-0068/21/04/381-386

[69] Saravanan R GRAIN software (private communication)

[70] Saravanan R, Kannan YB, Srinivasan N, Ismail I, (2017) J Supercond Nov Magn 30:407-419. https://doi.org/10.1007/s10948-016-3736-3

[71] Saravanan R, Saravanakumar S, Lavanya S (2010) Physica B 405:3700-3703. https://doi.org/10.1016/j.physb.2010.05.069

[72] Satalkar M, Kane SN (2016) J Phys: Conference Series 755:012050 DOI:10.1088/1742-6596/755/1/012050. https://doi.org/10.1088/1742-6596/755/1/012050

[73] Satishverma, Jagdishchand, Batoo KM, Singh M (2013) J Alloys Compd 551:715-721. https://doi.org/10.1016/j.jallcom.2012.10.163

[74] Seema J, Manoj K, Sandeep C, Geetika S, Mukesh J, Singh VN (2014) J Mole Struct 1076:55-62. https://doi.org/10.1016/j.molstruc.2014.07.048

[75] Shahbaz TF, Daadmebr V, Rozakhani AT, Hosseni AR, Gholipur S (2012) J Supercond Nov Magn 25(7):2445-2455.

[76] Sharma DR, Mathur R, Vadera SR, Kumar N, Kutty TRN (2003) J Alloys Compd 358:193-204. https://doi.org/10.1016/S0925-8388(03)00034-3

[77] Sheikh AD, Mathe VL (2008) J Mater Sci 43:2018-2025. https://doi.org/10.1007/s10853-007-2302-6

[78] Singh RK, Upadhyay C, Layek S, Yadav A (2010) Int J Eng Sci Technol 2(8):104-109.

[79] Slatineanu T, Iordan AR, Palamaru MN, Caltun OF, Gafton V, Leontie L (2011) Mater Res Bull 46:1455-1460. https://doi.org/10.1016/j.materresbull.2011.05.002

[80] Sozeri H, Durmus Z, Baykal A (2012) Mater Res Bull 47 2442-2448. https://doi.org/10.1016/j.materresbull.2012.05.036

[81] Sreeja V, Vijayanand S, Deka S, Joy PA (2008) Hyperfine Interact 183:99-107 https://doi.org/10.1007/s10751-008-9736-3

[82] Sujatha Ch, Venugopalreddy K, Sowri Babu K, Ramachandrareddy A, Rao KH (2013) Ceram Int 39:3077-3086. https://doi.org/10.1016/j.ceramint.2012.09.087

[83] Sutka A, Borisova A, Kleperis J, Mezinskis G, Jakovlevs D, Juhnevica I (2012) J Aus Ceram Soc 48(2): 150-155.

[84] Tehrani FS, Daadmehr V, Rezakhani AT, Akbarnejad RH, Gholipour S (2012) J Supercond Nov Magn 25(7):2443-2455. https://doi.org/10.1007/s10948-012-1655-5

[85] Trivedi UN, Chhantbar MC, Modi KB, Joshi HH (2005) Ind J Pure Appl Phys 43:688-690.

[86] Varalaxmi N, Sivakumar KV (2013) Mater Sci Eng C 33:145-152. https://doi.org/10.1016/j.msec.2012.08.022

[87] Varaprasad BBVS, Rajeshbabu R, Sivaramprasad M (2015) Mater Sci-Poland 33(4): 806-815.

[88] Verma A, Goel TC, Mendiratta RG, Alam MI (1999) Mater Sci Engg B 60:156-162. https://doi.org/10.1016/S0921-5107(99)00019-7

[89] Verma A, R. Chatterjee R (2006) J Magn Magn Mater 306:313-320. https://doi.org/10.1016/j.jmmm.2006.03.033

[90] Wagner KW (1913) Ann Phys 40:817-855. https://doi.org/10.1002/andp.19133450502

[91] Zaki HM, Al-Heniti SH, Elmosalami TA (2015) J Alloys Compd 633:104-114 https://doi.org/10.1016/j.jallcom.2015.01.304

[92] Zhang M, Zi Z, Liu Q, Zhang P, Tang X, Yang J, Zhu X, Sun Y,Dai J (2013) Adv Mater Sci Eng 609819(10 pages).

Chapter 4

Conclusions

Abstract

Conclusions of the findings on the synthesized samples are reported in this chapter.

Keywords

Sintered Samples, Structural Varation, Nickel Ferrite, Nickel-Zinc Ferrite, Cobalt-Zinc Ferrite, Manganese-Zinc Ferrite, Nickel-Copper-Zinc Ferrite, Magnesium-Copper-Zinc Ferrite

Contents

4   Summary 153

4.1 $Ni_{0.5}Zn_{0.5}Fe_2O_4$ samples sintered from 700-1400°C 153

4.2 $Ni_{0.8}Zn_{0.2}Fe_2O_4$ sample 155

4.3 $Mn_{0.4}Zn_{0.6}Fe_2O_4$ and $Co_{0.4}Zn_{0.6}Fe_2O_4$ samples 155

4.4 $Ni_{0.53}Cu_{0.12}Zn_{0.35}Fe_2O_4$ and $Mg_{0.2}Cu_{0.3}Zn_{0.5}Fe_2O_4$ samples 155

4.5 $NiFe_2O_4$ sample 156

4   Summary

Based on the results (Chapter − 2) and discussions (Chapter − 3) of the samples synthesized in the present research study, the important observations are given here as conclusion.

4.1 $Ni_{0.5}Zn_{0.5}Fe_2O_4$ samples sintered from 700-1400°C

Sintering effect on the $Ni_{0.5}Zn_{0.5}Fe_2O_4$ samples sintered in the temperature range 700-1400°C and prepared by mechanical alloying via high energy ball milling is consolidated below.

Cation distribution among the tetrahedral and octahedral sites is calculated theoretically and the agreement between observed and calculated X-ray intensity ratio confirms the calculated cation distribution and hence, the cation distribution is considered as occupancy parameter, for the respective samples, during Rietveld refinement. The agreement between the (i) theoretical and experimental lattice parameter (ii) observed and calculated structure factor and (iii) experimental and theoretical Bohr magneton values confirm the cation distribution in the present study.

Cation distribution indicates prevailing of mixed spinel structure in the unit cell. Additional phase is found in all samples. The occupation of zinc at octahedral (B) site decreases with increasing sintering temperature except samples sintered at 1100 and 1400°C whereas occupation of nickel at octahedral (B) site decreases upto a sintering temperature of 1000°C and thereafter increases with increase in sintering temperature. The experimental lattice parameter value is larger when compared with that of bulk sample. The experimental lattice parameter value decreases with increase in sintering temperature except for the samples sintered at 1200 and 1400°C. The difference between the experimental lattice parameter and theoretical lattice parameter decreases with increase in sintering temperature. The average crystallite size, which is evaluated using Scherrer formula, of the samples increases with increase in sintering temperature. Porosity decreases with increasing sintering temperature.

Bond strength between various sites in the unit cell is evaluated using maximum entropy method and the results are compared with earlier theoretical results. Tetrahedral – octahedral site interactions of the samples agrees well with the result of theoretical results except for samples sintered at 1100 and 1200°C. The sample sintered at 1200°C shows completely contradicted results in all interactions between various sites. Interestingly, tetrahedral - tetrahedral sites interactions are found to be the stronger than octahedral - octahedral sites interactions in all the samples. These contradictions are explained with proper mechanism.

Saturation magnetization is explained on the basis of cation distribution between tetrahedral and octahedral site. Yaffet-Kittel angle is present in the sample sintered at 700 and 1200°C. Ferromagnetic nature is very strong in sample sintered at 1300°C among the synthesized sample.

The optical band gap energy decreases up to a sintering temperature of 1000°C and thereafter the trend is reversed.

Dielectric properties are studied in $Ni_{0.5}Zn_{0.5}Fe_2O_4$ samples sintered from 700 to 1200°C. Dielectric constant decreases with increase in sintering temperature. The decrease in ε' is explained as a combined effect of number of ferrous ions on tetrahedral-site and thickness of grain boundary and besides these factors, presence of second phase also contributes to the high value of ε' in the case of sample sintered at 700°C. The dielectric constant and AC conductivity show normal dielectric behaviour with increase in sintering temperature and their behaviour is explained on the basis of hopping of charge carriers. The sample sintered at 1100°C shows low dielectric constant and high resistivity which is suitable for high frequency applications.

4.2 $Ni_{0.8}Zn_{0.2}Fe_2O_4$ sample

$Ni_{0.8}Zn_{0.2}Fe_2O_4$ nano ferrite particles were prepared by co-precipitation method. Spinel phase, along with a small additional peak corresponding to hematite phase, is confirmed from XRD data. The agreement between the (i) theoretical and experimental lattice parameter, (ii) observed and calculated X-ray intensity ratio, and (iii) experimental and theoretical Bohr magneton values confirm the cation distribution in the present study. The morphological study reveals the homogeneous nature of the sample. Absence of Yaffet-Kittel angle confirms the existence of collinear spin, which is further established by the mid bond electron density values between various atoms at A and B sites evaluated from maximum entropy method. The ferromagnetic nature of the sample is ascertained from the magnetic measurement.

4.3 $Mn_{0.4}Zn_{0.6}Fe_2O_4$ and $Co_{0.4}Zn_{0.6}Fe_2O_4$ samples

Manganese and cobalt substituted zinc ferrite ($Mn_{0.4}Zn_{0.6}Fe_2O_4$ and $Co_{0.4}Zn_{0.6}Fe_2O_4$) nano particles were synthesized successfully by co-precipitation method. The substitution causes the value of lattice parameter to decrease. The particle size of $Mn_{0.4}Zn_{0.6}Fe_2O_4$ ferrite is larger than that of $Co_{0.4}Zn_{0.6}Fe_2O_4$ ferrite due to the radii of substituting ions. Existence of ferrous ions in $Co_{0.4}Zn_{0.6}Fe_2O_4$ ferrite is attributed to the slight difference observed in the radii of tetrahedral and octahedral sites. MEM study reveals that tetrahedral (A) – octahedral (B) site interaction is the strongest interaction whereas magnetic study detected presence of Yaffet-Kittel angle. This contradiction is explained with the shrinkage in the lattice parameter value. Octahedral – octahedral site interaction in the unit cell is found to be the weakest interaction. Obtained optical band gap energy value is high due to the presence of second phase.

4.4 $Ni_{0.53}Cu_{0.12}Zn_{0.35}Fe_2O_4$ and $Mg_{0.2}Cu_{0.3}Zn_{0.5}Fe_2O_4$ samples

Copper and zinc substituted nickel ferrite ($Ni_{0.53}Cu_{0.12}Zn_{0.35}Fe_2O_4$) and magnesium and copper substituted zinc ferrite ($Mg_{0.2}Cu_{0.3}Zn_{0.5}Fe_2O_4$) have been synthesized by co-precipitation method. The variation in lattice parameter value of $Ni_{0.53}Cu_{0.12}Zn_{0.35}Fe_2O_4$ and $Mg_{0.2}Cu_{0.3}Zn_{0.5}Fe_2O_4$ samples evaluated from the Rietveld refinement method using

JANA (2006), is attributed to the radii of substituting ions. The doping enhances the density in both samples. Magnetic studies reveal the weakening of tetrahedral – octahedral site interactions in the unit cell due to the presence of Yaffet-Kittel angles whereas maximum entropy method analysis shows that tetrahedral – octahedral site interaction in the unit cell is the strongest interaction in both samples. Disorder in cation distribution, fine particle size and the increase in magnetic saturation values were attributed to the contradicted results obtained from magnetic and maximum entropy method studies. $Ni_{0.53}Cu_{0.12}Zn_{0.35}Fe_2O_4$ sample shows ferromagnetic nature whereas $Mg_{0.2}Cu_{0.3}Zn_{0.5}Fe_2O_4$ shows paramagnetic nature.

4.5 $NiFe_2O_4$ sample

XRD data of the nickel ferrite sample prepared using solid state reaction method, is refined using Rietveld refinement method. The agreement between the theoretical and experimental lattice parameter value confirms the cation distribution. The experimental lattice parameter value is correlated with the number of Fe^{2+} ions at octahedral site. Nano meter range particles size is confirmed from the morphology study.   XRF characterization shows the purity of the sample. The mid bond electron density values between various atoms at tetrahedral and octahedral sites is found out using maximum entropy method. Tetrahedral – tetrahedral site interaction in the unit cell is found to be strongest among the three interactions in the present study. The magnetic measurement revealed that the sample has low value of saturation magnetization. High value (10^6) of dielectric constant is observed which is attributed to more number of Fe^{2+} ions at octahedral site.

Keyword Index

Basic Concepts 1

Cation Distribution 120

Characterization Techniques 1

$Co_{0.4}Zn_{0.6}Fe_2O_4$ 1

Cobalt-Zinc Ferrite 153

Dielectric Properties 120

Dielectric 42

DTA 42

EDX 42

Electron Density 42

Ferrite materials 1

Magnesium-Copper-Zinc Ferrite 153

Magnetic Properties 120

Magnetic 1

Manganese-Zinc Ferrite 153

$Mg_{0.2}Cu_{0.3}Zn_{0.5}Fe_2O_4$ 1

$Mn_{0.4}Zn_{0.6}Fe_2O_4$ 1

$Ni_{0.53}Cu_{0.12}Zn_{0.35}Fe_2O_4$ 1

$Ni_{0.8}Zn_{0.2}Fe_2O_4$ 1

Nickel Ferrite 153

Nickel-Copper-Zinc Ferrite 153

Nickel-Zinc Ferrite 153

$NiFe_2O_4$ 1

Optical Properties 120

Powder XRD 42

SEM 42

Sintered Samples 153

Structural Parameters 120

Structural Varation 153

TG/DTA 120

TGA 42

X-ray Fluorescence 42

About the Author

Dr Ramachandran Saravanan, has been associated with the Department of Physics, The Madura College, affiliated with the Madurai Kamaraj University, Madurai, Tamil Nadu, India from the year 2000. He is the head of the Research Centre and PG department of Physics. He worked as a research associate during 1998 at the Institute of Materials Research, Tohoku University, Sendai, Japan and then as a visiting researcher at Centre for Interdisciplinary Research, Tohoku University, Sendai, Japan up to 2000.

Earlier, he was awarded the Senior Research Fellowship by CSIR, New Delhi, India, during Mar. 1991 - Feb.1993; awarded Research Associateship by CSIR, New Delhi, during 1994 – 1997. Then, he was awarded a Research Associateship again by CSIR, New Delhi, during 1997- 1998. Later he was awarded the Matsumae International Foundation Fellowship in1998 (Japan) for doing research at a Japanese Research Institute (not availed by him due to the simultaneous occurrence of other Japanese employment).

He has guided eleven Ph.D. scholars as of 2017, and about five researchers are working under his guidance on various research topics in materials science, crystallography and condensed matter physics. He has published around 140 research articles in reputed Journals, mostly International, apart from around 50 presentations in conferences, seminars and symposia. He has also guided around 60 M.Phil. scholars and an equal number of PG students for their projects. He has attracted government funding in India, in the form of Research Projects. He has completed two CSIR (Council of Scientific and Industrial Research, Govt. of India), one UGC (University Grants Commission, India) and one DRDO (Defense Research and Development Organization, India) research projects successfully and is proposing various projects to Government funding agencies like CSIR, UGC and DST.

He has written 8 books in the form of research monographs including; "Experimental Charge Density - Semiconductors, oxides and fluorides" (ISBN-13: 978-3-8383-8816-8; ISBN-10:3-8383-8816-X), "Experimental Charge Density - Dilute Magnetic Semiconducting (DMS) materials" (ISBN-13: 978-3-8383-9666-8; ISBN-10: 3-8383-9666-9) and "Metal and Alloy Bonding - An Experimental Analysis" (ISBN -13: 978-1-4471-2203-6). He has committed to write several books in the near future.

His expertise includes various experimental activities in crystal growth, materials science, crystallographic, condensed matter physics techniques and tools as in slow evaporation, gel, high temperature melt growth, Bridgman methods, CZ Growth, high vacuum sealing etc. He and his group are familiar with various equipment such as: different types of cameras; Laue, oscillation, powder, precession cameras; Manual 4-

circle X-ray diffractometer, Rigaku 4-circle automatic single crystal diffractometer, AFC-5R and AFC-7R automatic single crystal diffractometers, CAD-4 automatic single crystal diffractometer, crystal pulling instruments, and other crystallographic, material science related instruments. He and his group have sound computational capabilities on different types of computers such as: IBM – PC, Cyber180/830A – Mainframe, SX-4 Supercomputing system – Mainframe. He is familiar with various kind of software related to crystallography and materials science. He has written many computer software programs himself as well. Around twenty of his programs (both DOS and GUI versions) have been included in the SINCRIS software database of the International Union of Crystallography.

www.ingramcontent.com/pod-product-compliance
Lightning Source LLC
Chambersburg PA
CBHW071642210326
41597CB00017B/2077